The Dynamics of Modulated Wave Trains

MEMOIRS
of the
American Mathematical Society

Number 934

The Dynamics of Modulated
Wave Trains

Arjen Doelman
Björn Sandstede
Arnd Scheel
Guido Schneider

May 2009 • Volume 199 • Number 934 (fifth of 6 numbers) • ISSN 0065-9266

American Mathematical Society
Providence, Rhode Island

2000 *Mathematics Subject Classification.* Primary 35K57, 35A35, 35Q53, 37L99.

Library of Congress Cataloging-in-Publication Data

The dynamics of modulated wave trains / Arjen Doelman ... [et al.].
    p. cm. — (Memoirs of the American Mathematical Society, ISSN 0065-9266 ; no. 934)
"Volume 199, Number 934 (fifth of 6 numbers)."
Includes bibliographical references.
ISBN 978-0-8218-4293-5 (alk. paper)
    1. Reaction-diffusion equations. 2. Approximation theory. 3. Burgers equation. I. Doelman, A.
QA377.D86 2009
515'.3534—dc22
    2008055480

Memoirs of the American Mathematical Society

This journal is devoted entirely to research in pure and applied mathematics.

Subscription information. The 2009 subscription begins with volume 197 and consists of six mailings, each containing one or more numbers. Subscription prices for 2009 are US$709 list, US$567 institutional member. A late charge of 10% of the subscription price will be imposed on orders received from nonmembers after January 1 of the subscription year. Subscribers outside the United States and India must pay a postage surcharge of US$65; subscribers in India must pay a postage surcharge of US$95. Expedited delivery to destinations in North America US$57; elsewhere US$160. Each number may be ordered separately; *please specify number* when ordering an individual number. For prices and titles of recently released numbers, see the New Publications sections of the *Notices of the American Mathematical Society*.

Back number information. For back issues see the *AMS Catalog of Publications*.

Subscriptions and orders should be addressed to the American Mathematical Society, P. O. Box 845904, Boston, MA 02284-5904, USA. *All orders must be accompanied by payment.* Other correspondence should be addressed to 201 Charles Street, Providence, RI 02904-2294, USA.

Copying and reprinting. Individual readers of this publication, and nonprofit libraries acting for them, are permitted to make fair use of the material, such as to copy a chapter for use in teaching or research. Permission is granted to quote brief passages from this publication in reviews, provided the customary acknowledgment of the source is given.

Republication, systematic copying, or multiple reproduction of any material in this publication is permitted only under license from the American Mathematical Society. Requests for such permission should be addressed to the Acquisitions Department, American Mathematical Society, 201 Charles Street, Providence, Rhode Island 02904-2294, USA. Requests can also be made by e-mail to reprint-permission@ams.org.

Memoirs of the American Mathematical Society (ISSN 0065-9266) is published bimonthly (each volume consisting usually of more than one number) by the American Mathematical Society at 201 Charles Street, Providence, RI 02904-2294, USA. Periodicals postage paid at Providence, RI. Postmaster: Send address changes to Memoirs, American Mathematical Society, 201 Charles Street, Providence, RI 02904-2294, USA.

© 2009 by the American Mathematical Society. All rights reserved.
Copyright of this publication reverts to the public domain 28 years
after publication. Contact the AMS for copyright status.
This publication is indexed in *Science Citation Index*®, *SciSearch*®, *Research Alert*®,
CompuMath Citation Index®, *Current Contents*®/*Physical, Chemical & Earth Sciences*.
Printed in the United States of America.

∞ The paper used in this book is acid-free and falls within the guidelines
established to ensure permanence and durability.
Visit the AMS home page at http://www.ams.org/

10 9 8 7 6 5 4 3 2 1   14 13 12 11 10 09

Contents

Notation	1
Chapter 1. Introduction	3
1.1. Grasshopper's guide	3
1.2. Slowly-varying modulations of nonlinear wave trains	4
1.3. Predictions from the Burgers equation	7
1.4. Verifying the predictions made from the Burgers equation	8
1.5. Related modulation equations	12
1.6. References to related works	13
Chapter 2. The Burgers equation	15
2.1. Decay estimates	15
2.2. Fronts in the Burgers equation	17
Chapter 3. The complex cubic Ginzburg–Landau equation	19
3.1. Set-up	19
3.2. Slowly-varying modulations of the $k = 0$ wave train: Results	20
3.3. Derivation of the Burgers equation	23
3.4. The construction of higher-order approximations	24
3.5. The approximation theorem for the wave numbers	25
3.6. Mode filters, and separation into critical and noncritical modes	25
3.7. Estimates of the linear semigroups	29
3.8. Estimates of the residual	30
3.9. Estimates of the errors	31
3.10. Proofs of the theorems from §3.2	34
Chapter 4. Reaction-diffusion equations: Set-up and results	39
4.1. The abstract set-up	39
4.2. Expansions of the linear and nonlinear dispersion relations	41
4.3. Formal derivation of the Burgers equation	43
4.4. Validity of the Burgers equation	45
4.5. Existence and stability of weak shocks	48
Chapter 5. Validity of the Burgers equation in reaction-diffusion equations	53
5.1. From phases to wave numbers	53
5.2. Bloch-wave analysis	55
5.3. Mode filters, and separation into critical and noncritical modes	58
5.4. Estimates for residuals and errors	61
5.5. Proofs of the theorems from §4.4	63

Chapter 6. Validity of the inviscid Burgers equation in reaction-diffusion
          systems   65
  6.1. An illustration: The Ginzburg–Landau equation   65
  6.2. Formal derivation of the conservation law   66
  6.3. Validity of the inviscid Burgers equation   67
  6.4. Proof of the theorems from §6.3   68

Chapter 7. Modulations of wave trains near sideband instabilities   73
  7.1. Introduction   73
  7.2. An illustration: The Ginzburg–Landau equation   74
  7.3. Validity of the Korteweg–de Vries and the Kuramoto–Sivashinsky
          equation   75
  7.4. Proof of Theorem 7.2   78
  7.5. Proof of Theorem 7.5   79

Chapter 8. Existence and stability of weak shocks   83
  8.1. Proof of Theorem 4.10   83
  8.2. Proof of Theorem 4.12   88

Chapter 9. Existence of shocks in the long-wavelength limit   93
  9.1. A lattice model for weakly interacting pulses   93
  9.2. Proof of Theorem 9.2   95

Chapter 10. Applications   99
  10.1. The FitzHugh–Nagumo equation   99
  10.2. The weakly unstable Taylor–Couette problem   100

Bibliography   103

Abstract

We investigate the dynamics of weakly-modulated nonlinear wave trains. For reaction-diffusion systems and for the complex Ginzburg–Landau equation, we establish rigorously that slowly varying modulations of wave trains are well approximated by solutions to the Burgers equation over the natural time scale. In addition to the validity of the Burgers equation, we show that the viscous shock profiles in the Burgers equation for the wave number can be found as genuine modulated waves in the underlying reaction-diffusion system. In other words, we establish the existence and stability of waves that are time-periodic in appropriately moving coordinate frames which separate regions in physical space that are occupied by wave trains of different, but almost identical, wave number. The speed of these shocks is determined by the Rankine–Hugoniot condition where the flux is given by the nonlinear dispersion relation of the wave trains. The group velocities of the wave trains in a frame moving with the interface are directed toward the interface. Using pulse-interaction theory, we also consider similar shock profiles for wave trains with large wave number, that is, for an infinite sequence of widely separated pulses. The results presented here are applied to the FitzHugh–Nagumo equation and to hydrodynamic stability problems.

Received by the editor 10 January 2005.

2000 *Mathematics Subject Classification*. Primary 35K57, 35A35, 35Q53, 37L99.

Key words and phrases. Approximation, wave trains, modulated waves, phase diffusion, viscous Burgers equation.

This work was supported by the Volkswagenstiftung through the *Research in Pairs* (RiP) program at the Mathematisches Forschungsinstitut Oberwolfach in September 2000. We are grateful to Thierry Gallay and Hannes Uecker for many comments that helped us to improve the paper.

B. Sandstede was partially supported by an Alfred P Sloan Research Fellowship, by a Royal Society Wolfson Research Merit Award, and the NSF under grants DMS-9971703 and DMS-0203854.

A. Scheel was partially supported by the NSF through grant DMS-0203301.

G. Schneider would like to thank Ian Melbourne for stimulating discussions. G. Schneider was partially supported by the Deutsche Forschungsgemeinschaft DFG under the grant Kr 690/18-1/2.

Notation

Throughout this paper, we denote possibly different constants by the same symbol C. We denote by H^m_{ul} the space of locally square-integrable functions on $\mathbb{R}$ whose first m weak derivatives exist and are uniformly bounded in local L^2 spaces and for which the spatial translation $y \mapsto u(\cdot + y)$ is continuous with respect to the H^m_{ul}-norm. Their norm is defined by

$$\|u\|_{H^m_{\mathrm{ul}}} = \sup_{x \in \mathbb{R}} \|u\|_{H^m(x, x+1)}$$

where the Sobolev norm $\|\cdot\|_{H^m(x, x+1)}$ is, for each fixed $x \in \mathbb{R}$, given by

$$\|u\|_{H^m(x, x+1)} = \sum_{j=0}^{m} \|\partial_x^j u\|_{L^2(x, x+1)}.$$

We also use the weighted Sobolev spaces $H^m(n)$ which we equip with the norm

$$\|u\|_{H^m(n)} = \|u \rho_{\mathrm{w}}^n\|_{H^m}$$

where $\rho_{\mathrm{w}}(x) = \sqrt{1 + |x|^2}$.

$u(x, t)$	solution to reaction-diffusion system
$u_0(\omega t - kx; k)$	wave train (2π-periodic in argument θ)
$\theta = \omega t - kx$	travelling-wave coordinate (wave train)
k	wave number
ω	temporal frequency
$\omega_{\mathrm{nl}}(k)$	nonlinear dispersion relation
$c_{\mathrm{p}} = \omega_{\mathrm{nl}}(k)/k$	phase velocity
$c_{\mathrm{g}} = \mathrm{d}\omega_{\mathrm{nl}}(k)/\mathrm{d}k$	group velocity
λ	temporal eigenvalue
$\lambda_{\mathrm{lin}}(\nu)$	linear dispersion relation in comoving frame
ν	complex spatial Floquet exponent
ℓ	imaginary part of spatial Floquet exponent $\nu = i\ell$
$\Phi(X, T)$	slowly varying phase
$q(X, T)$	slowly varying wave number
$0 < \delta \ll 1$	multi-scale expansion parameter
$(X, T) = (\delta(x - c_{\mathrm{g}} t), \delta^2 t)$	slow space and time variables
c_*	speed of viscous shock
ω_*	frequency of viscous shock
$\xi = x - c_* t$	travelling-wave coordinate (shock)
$\tau = \omega_* t$	rescaled time (2π-periodic)
$\sigma = \tau - k_0 \xi$	co-rotating coordinate (wave train)
$\hat{u}(\ell) = [\mathcal{F} u](\ell)$	Fourier transform of $u(x)$
$\check{u}(x, \ell) = [\mathcal{J} u](x, \ell)$	Bloch transform of $u(x)$

When we fix a wave number k_0, we will denote the associated frequency, group and phase velocities evaluated at k_0 by $\omega_0 = \omega_{\mathrm{nl}}(k_0)$, c_{p}^0 and c_{g}^0, respectively. When confusion is unlikely, we will drop the index 0.

CHAPTER 1

Introduction

We begin in §1.1 with a grasshopper's guide which contains a brief outline of our results and the plan of the paper. In the rest of the introduction, starting with §1.2, we explain our results and their proofs in more detail. We finish the introduction in §1.6 with references to related work and a brief discussion of open problems.

1.1. Grasshopper's guide

The issue investigated in this paper is the dynamics of slow modulations of nonlinear spatially-periodic travelling waves, in the following referred to as wave trains, in reaction-diffusion equations

$$\partial_t u = D\partial_{xx} u + f(u), \qquad x \in \mathbb{R}, \quad u \in \mathbb{R}^d.$$

Let $u(x,t) = u_0(\omega t - kx; k)$ be such a wave train whose profile $u_0(\theta;k)$ is 2π-periodic in θ, and whose temporal frequency ω and spatial wave number k are related through the nonlinear dispersion relation $\omega = \omega_{\mathrm{nl}}(k)$. We define the group velocity of the wave trains to be $c_{\mathrm{g}} = \omega'_{\mathrm{nl}}(k)$ and denote their linear dispersion relation by $\lambda_{\mathrm{lin}}(\nu)$. If we modulate the wave number k of the wave trains over large spatial scales, we are led to an ansatz of the form

$$u(x,t) = u_0(\omega t - kx - \Phi(X,T); k + \delta\partial_X \Phi(X,T)), \qquad (X,T) = \big(\delta(x - c_{\mathrm{g}}t), \delta^2 t\big)$$

with $0 < \delta \ll 1$, which turns out to satisfy the underlying reaction-diffusion system formally to leading order provided the wave-number modulation $q(X,T) = \partial_X \Phi(X,T)$ is a solution of the (viscous) Burgers equation

$$\partial_T q = \frac{1}{2}\lambda''_{\mathrm{lin}}(0)\partial_{XX} q - \frac{1}{2}\omega''_{\mathrm{nl}}(k)\partial_X(q^2).$$

In this manuscript, we investigate the following issues:

- *Validity results for the Burgers equation (setup: §4.1; results: §4.4; proofs: §5):*
  We establish rigorous error estimates for the approximation of slowly-varying modulated wave trains via the Burgers equation over the natural time scale of order δ^{-2}. The error estimates are uniform in the spatial variable x provided the wave-number modulation $q(X,T)$ approaches limits as $X \to \pm\infty$. These results are formulated and proved separately for the complex Ginzburg–Landau equation (§3) and for general reaction-diffusion systems. For the latter case, we also present approximation results for the inviscid Burgers equation over time scales of order δ^{-1} (§6).

- *Modulation equations near sideband instabilities (§7):*
  When the underlying wave trains become sideband unstable, the Burgers equation does no longer provide an accurate description of the dynamics of slow modulations. Instead, depending on the form of the linear dispersion relation, it is the Korteweg–de Vries or the Kuramoto–Sivashinsky equation that takes its role. We discuss their validity properties for reaction-diffusion systems.
- *Existence and stability of weak shocks (setup: §4.1; results: §4.5; proofs: §8):*
  We show that the viscous shock fronts in the Burgers equation correspond to genuine modulated waves of the underlying reaction-diffusion system. In other words, we construct stable waves that are time-periodic in an appropriately moving coordinate frame and whose profile converges, as $x \to \pm\infty$, to two wave trains with different, but almost identical, wave number. The speed of these interfaces is determined by the Rankine–Hugoniot condition with the flux given by the nonlinear dispersion relation of the wave trains. The group velocities of the asymptotic wave trains, computed in a frame moving with the interface, are directed toward the interface.
- *Global analysis of trains of well-separated pulses (§9):*
  In the limit of infinite wavelength (or zero wave number), wave trains are made up by an infinite number of well-separated pulses whose dynamics can be described formally by a lattice dynamical system. In this description, modulated fronts that connect two such wave trains can be found as heteroclinic orbits to a certain delay equation. We show that these heteroclinic solutions exist between any two wave trains.
- *Applications (§10):*
  The results presented here are applied to the FitzHugh–Nagumo equation and to the Taylor–Couette problem.

1.2. Slowly-varying modulations of nonlinear wave trains

We shall investigate the dynamics of weakly-modulated nonlinear wave trains in partial differential equations (PDEs) on the real line. To set the scene, suppose that we are given a reaction-diffusion system

$$(1.1) \qquad \partial_t u = D \partial_{xx} u + f(u), \qquad x \in \mathbb{R}, \quad u \in \mathbb{R}^d.$$

Starting point of our investigation are wave trains which are solutions $u(x,t) = u_0(\omega t - kx)$ of (1.1) that are 2π-periodic in their argument $\theta = \omega t - kx$. Thus, ω may be interpreted as the temporal frequency of the wave train and k as its spatial wave number; their quotient $c_{\mathrm{p}} = \omega/k$ gives the wave speed, or phase velocity, of the nonlinear wave. Typically, wave trains exist for an entire range[1] of wave numbers k, and both the profile $u_0(\theta)$ and the frequency ω will depend on the choice of k. To reflect this fact, we write the travelling wave as

$$(1.2) \qquad u(x,t) = u_0(\omega t - kx; k)$$

and denote the frequency ω selected by the wave number k by $\omega_{\mathrm{nl}}(k)$; we shall refer to this function as the nonlinear dispersion relation.

---

[1] See §4.1 for details.

1.2. SLOWLY-VARYING MODULATIONS OF NONLINEAR WAVE TRAINS

We will assume that the wave trains are spectrally stable: If we transform (1.1) into the frame $\theta = \omega t - kx$ and linearize the resulting equation about the wave trains $u_0(\theta; k)$, we obtain the linear operator

(1.3) $$\mathcal{L}u = k^2 D \partial_{\theta\theta} u - \omega \partial_\theta u + f'(u_0(\theta; k))u.$$

Since the coefficients appearing in (1.3) are 2π-periodic in θ, a complex number $\lambda \in \mathbb{C}$ is in the spectrum of $\mathcal{L}$, considered for instance on $L^2(\mathbb{R})$, if, and only if, there is a wave number $\ell \in \mathbb{R}$ and a nonzero 2π-periodic function $v(\theta; \ell)$ such that

$$\mathcal{L}u = \lambda u, \qquad u(\theta) := e^{i\ell\theta} v(\theta; \ell).$$

Elements λ of the spectrum of $\mathcal{L}$ come in curves $\lambda(i\ell)$ that are parameterized by $\ell \in \mathbb{R}$. In particular, translation invariance of (1.1) implies[2] that there is a spectral curve $\lambda_{\text{lin}}(i\ell)$ with $\lambda_{\text{lin}}(0) = 0$; the associated eigenfunctions $v(\theta; \ell)$ have the expansion

$$v(\theta; \ell) = \partial_\theta u_0(\theta; k) - \ell \partial_k u_0(\theta; k) + O(\ell^2).$$

Spectral stability of the wave train therefore means that the spectrum of $\mathcal{L}$ lies in the open left half-plane except for the curve $\lambda_{\text{lin}}(i\ell)$ for which we assume that $\lambda_{\text{lin}}''(0) > 0$.

The fact that wave trains exist for wave numbers k in an open, nonempty interval of real numbers has interesting implications. We may, for instance, pick two wave numbers k_- and k_+ from the admissible range and prepare an initial condition which coincides with $u_0(-k_- x; k_-)$ on $\mathbb{R}^-$ and with $u_0(-k_+ x; k_+)$ on $\mathbb{R}^+$, perhaps with some smooth transition between the two patterns near $x = 0$. The dynamical behaviour of the resulting solution will reflect the interaction properties of the two wave trains with wave numbers k_- and k_+. Worded differently, the region where the transition from wave number k_- to k_+ occurs can be thought of as an interface between the two patterns described by the chosen wave trains. Of interest is then the dynamics of this interface.

More generally, we are interested in the dynamics of modulated wave trains. To motivate the ansatz in which we shall seek solutions, we begin with a brief heuristic discussion of local wave numbers; Figure 1.1 contains a graphical illustration. Since $u_0(\theta)$ is 2π-periodic in its argument, we can interpret the number k in the wave train $u_0(-kx)$ as the number of waves per unit interval: its wave number, in other words. For functions of the form $u_0(-\phi(x))$, we may therefore regard the derivative $\partial_x \phi(x)$ as the local wave number of $u_0(-\phi(x))$ near x. This concept becomes more plausible and credible if $\partial_x \phi$ varies very little in x. Thus, pick a function $\Phi(X)$ and choose a small multi-scale parameter δ with $0 < \delta \ll 1$. The function $u_0(-kx - \Phi(\delta x))$ has indeed, locally at x, a wave number that is given by $k + \delta \partial_X \Phi(\delta x)$. This interpretation may become more clear when specializing to $\Phi(X) = X$ in which case $u_0(-kx - \Phi(\delta x)) = u_0(-(k+\delta)x)$ so that local and global wave numbers coincide.

Thus, exploiting the freedom we have in selecting the wave number, we choose a smooth real-valued function $\Phi(X)$ and consider a slowly modulated wave train of the form

(1.4) $$u(x) = u_0(-kx - \Phi(\delta x); k + \delta \partial_X \Phi(\delta x))$$

for $0 < \delta \ll 1$, where we think of δ as a small parameter that determines the length scale over which the wave number is modulated by the function $\partial_X \Phi$. For

---

[2]See §4.1 for details.

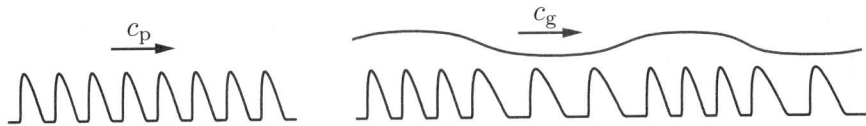

FIGURE 1.1. The left panel shows a wave train that travels with its phase velocity c_p, while the right panel contains a modulated wave train with its local wave number q shown on top. The slowly-varying modulation of the wave number propagates with the group velocity c_g.

$\delta > 0$ sufficiently small, it is reasonable to expect that the solution of (1.1) with initial condition (1.4) remains a slowly modulated wave train: it should still be of the form (1.4) for a function $\Phi(X)$ that now also depends on time. If this is true, then it ought to be possible to derive an effective evolution equation for the time-dependent modulation Φ.

Evolution equations that describe slowly-modulated wave trains have indeed been derived in the literature, starting with the seminal work [23], and we shall now describe the outcome of these analyses. First, we make the modulation ansatz

$$(1.5) \qquad u(x,t) = u_0(\omega_\mathrm{nl}(k)t - kx - \Phi(X,T); k + \delta\partial_X\Phi(X,T))$$

for $0 < \delta \ll 1$. The variables (X,T) represent the relevant length and time scales over which the slowly-varying modulation of the wave number evolves. The correct choice for (X,T) turns out to be

$$(1.6) \qquad (X,T) = \bigl(\delta(x - c_\mathrm{g}t), \delta^2 t\bigr),$$

where

$$(1.7) \qquad c_\mathrm{g} = \omega'_\mathrm{nl}(k)$$

is what we shall refer to as the group velocity of the wave trains. In these coordinates, the local wave number $q(X,T) := \partial_X\Phi(X,T)$ satisfies, to leading order and on a formal level, the Burgers equation[3]

$$(1.8) \qquad \partial_T q = \frac{1}{2}\lambda''_\mathrm{lin}(0)\partial_{XX}q - \frac{1}{2}\omega''_\mathrm{nl}(k)\partial_X(q^2),$$

while the phase $\Phi(X,T)$ itself satisfies the integrated Burgers equation

$$(1.9) \qquad \partial_T\Phi = \frac{1}{2}\lambda''_\mathrm{lin}(0)\partial_{XX}\Phi - \frac{1}{2}\omega''_\mathrm{nl}(k)(\partial_X\Phi)^2.$$

At this point, two questions arise naturally:
- First, assuming that the description via (1.8) is correct, what do we learn from it with regard to the dynamics of modulated wave trains of the reaction-diffusion equation (1.1)?
- Second, can we prove that (1.5) and (1.8) together describe the dynamics of modulated wave trains accurately? In other words, given a solution $q(X,T)$ to the Burgers equation, is there a solution to the reaction-diffusion equation (1.1) which differs from (1.5) by terms that go to zero sufficiently fast as $\delta \to 0$?

---

[3]The term *Burgers equation* is often reserved for the *inviscid* Burgers equation without the diffusion term in (1.8). We break with this convention and shall refer to (1.8) as the Burgers equation.

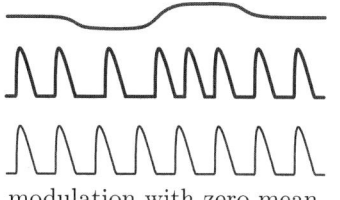 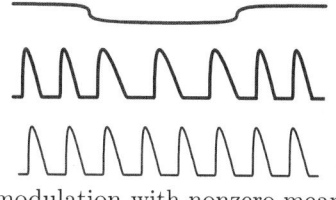

FIGURE 1.2. The difference between localized modulations with zero (left) and nonzero (right) mean is illustrated. The upper row shows the local wave number as a function of x, the resulting modulated wave is plotted in the center row, and the original wave train is shown in the third row for comparison. The phases of the modulated wave and the wave train at $x = \pm\infty$ coincide for wave-number modulations with zero mean, while modulations with nonzero mean generate a phase difference equal to the mean.

1.3. Predictions from the Burgers equation

To address the first question (whose answer will motivate why we may want to look into the second issue), we briefly review[4] some key features of solutions to the Burgers equation. First, we record that solutions $q(X, T)$ of the Burgers equation with sufficiently localized initial data decay to zero in $L^1 \cap L^\infty$ as $T \to \infty$. In terms of (1.1), this means that localized perturbations of the wave number of the wave trains will decay to zero.

Localized solutions $\Phi(X, T)$ of (1.9), which correspond to solutions $q = \partial_X \Phi$ of (1.8) with vanishing spatial mean, look very much like a Gaussian when renormalized appropriately and, in particular, remain localized near $X = 0$ for all T. Taking into account the frame (1.6) in which the Burgers equation has been derived, we see that sufficiently localized perturbations of the phase of wave trains decay to zero in $L^1 \cap L^\infty$ while being localized near $x = c_{\mathrm{g}} t$ when suitably renormalized. This justifies the term *group velocity* for the quantity c_{g}: it is the velocity with which localized perturbations in phase or wave number propagate in time (see also Figure 1.1). Localization of the wave number $q(X, T)$ simply means that the wave trains at $X = \pm\infty$ are the same. The phase $\Phi(X, T)$ is localized precisely when $q(X, T)$ has vanishing spatial mean:

$$\Phi(X, T) = \int_{-\infty}^{X} q(Y, T)\, \mathrm{d}Y, \qquad \Phi(\infty, T) - \Phi(-\infty, T) = \int_{-\infty}^{\infty} q(X, T)\, \mathrm{d}X.$$

Localization of the phase function $\Phi(X, T)$ therefore means that the wave trains at $X = \pm\infty$ are in phase, and the effect of (1.5) is simply to move some of the interior waves forth and back. We refer to Figure 1.2 for an illustration.

Next, we shall focus on nonlocalized solutions of the Burgers equation. Of particular interest are the viscous Lax shocks of (1.8). For any two given numbers $q_-, q_+ \in \mathbb{R}$, the Burgers equation admits a travelling wave $q(X, T) = q_*(X - c^*T)$ with $q_*(X - c^*T) \to q_\pm$ as $X \to \pm\infty$ if, and only if,

(1.10) $$\omega_{\mathrm{nl}}''(k)(q_+ - q_-) < 0.$$

---

[4]Details can be found in §2.

Thus, $q_- < q_+$ for $\omega_{\text{nl}}''(k) < 0$ and $q_+ < q_-$ for $\omega_{\text{nl}}''(k) > 0$. If (1.10) is met, then the wave speed c^* is given by

$$\text{(1.11)} \qquad c^* = \frac{1}{2}\omega_{\text{nl}}''(k)(q_+ + q_-),$$

and the front $q_*(X - c^*T)$ is asymptotically stable. The Burgers equation (1.8) is a viscous conservation law, and the inverse slopes of its characteristics to either side of the front q_* are given by

$$c_{\text{g}}^{\pm} = \omega_{\text{nl}}''(k)q_{\pm}.$$

In particular, using (1.10) and (1.11), we have

$$\text{(1.12)} \qquad c_{\text{g}}^- > c^* > c_{\text{g}}^+,$$

and the characteristics therefore point toward the front interface which confirms that the fronts q_* are viscous Lax shocks. Of course, if (1.10) is not met, then the characteristics point away from the transition area near $X = 0$, and we expect the solution $q(X, T)$ of (1.8) to behave in a way that is similar to rarefaction waves in conservation laws.

Starting from a wave train with wave number k in the reaction–diffusion system, the Lax shocks $q_*(X - c^*T)$ of the associated Burgers equation correspond formally, via (1.5), to coherent structures of the reaction-diffusion system (1.1) that move with speed $c_* = c_{\text{g}} + \delta c^*$ and connect the wave train with wave number $k_- = k + \delta q_-$ at $X = -\infty$ to the wave train with wave number $k_+ = k + \delta q_+$ at $X = \infty$ (see Figure 1.3). Since the group velocities of the asymptotic wave trains are given by

$$\omega_{\text{nl}}'(k + \delta q_{\pm}) = c_{\text{g}} + \omega_{\text{nl}}''(k)\delta q_{\pm} = c_{\text{g}} + \delta c_{\text{g}}^{\pm},$$

we can therefore conclude from (1.12) that

$$\omega_{\text{nl}}'(k + \delta q_-) > c_{\text{g}} + \delta c^* > \omega_{\text{nl}}'(k + \delta q_+).$$

Using the definitions $k_{\pm} = k + \delta q_{\pm}$ and $c_* = c_{\text{g}} + \delta c^*$, we finally get

$$\omega_{\text{nl}}'(k_-) > c_* > \omega_{\text{nl}}'(k_+).$$

Using the interpretation of the group velocity as the speed with which perturbation propagate along the wave train, we therefore see that the asymptotic wave trains transport perturbations toward the interface that separates them since its speed is $c_* = c_{\text{g}} + \delta c^*$. Coherent structures with this property are often referred to as sinks [50].

This ends our discussion of the dynamics of the Burgers equation and how it relates to the underlying reaction-diffusion equation. In the next section, we outline our approach for proving validity of the Burgers equation as the system that governs modulations of nonlinear wave trains and the existence of the Lax shocks we discussed above in the reaction-diffusion equation.

1.4. Verifying the predictions made from the Burgers equation

We will prove in §8 that the coherent structures which we discussed in the previous section indeed exist. More precisely, for given constants k_- and k_+, we are interested in solutions to (1.1) of the form $u(x, t) = u_*(x - c_*t, \omega_*t)$, where $u_*(\xi, \tau)$ is 2π-periodic in τ and

$$u_*(x - c_*t, \omega_*t) \to u_0(\omega_{\pm}t - k_{\pm}x - \phi_{\pm}; k_{\pm})$$

1.4. VERIFYING THE PREDICTIONS MADE FROM THE BURGERS EQUATION

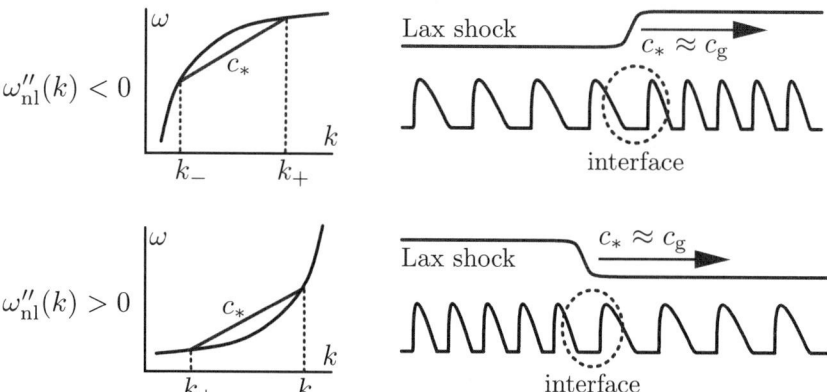

FIGURE 1.3. The relation between concave (top) and convex (bottom) dispersion relations and the resulting weak shock profiles is illustrated. The velocity c_* of the interface, given by the Rankine–Hugoniot condition (1.13) and therefore by the slopes labelled c_* in the insets, is close to the group velocity $c_g = \omega'_{nl}$, which is positive for the dispersion relations shown to the left.

as $x \to \pm\infty$ for appropriate phase constants $\phi_\pm \in \mathbb{R}$. Solutions $u_*(x - c_* t, \omega_* t)$ of the above form are spatially asymptotic to the wave trains with wave numbers k_- and k_+. They are also temporally periodic with frequency ω_* when considered in a coordinate frame that moves with speed c_*. It turns out that the assumption of temporal periodicity determines the wave speed via the Rankine–Hugoniot condition

$$(1.13) \qquad c_* = \frac{\omega_{nl}(k_+) - \omega_{nl}(k_-)}{k_+ - k_-},$$

which coincides with the speed prediction (1.11) once the moving frame (1.6) is taken into account. The temporal frequency ω_* is given by

$$\omega_* = \frac{\omega_{nl}(k_-)k_+ - \omega_{nl}(k_+)k_-}{k_+ - k_-}.$$

We shall show that u_* exists, and is nonlinearly stable with respect to (1.1), whenever the wave numbers $k_\pm$ are sufficiently close to each other[5] and the Lax condition

$$\omega'_{nl}(k_-) > c_* > \omega'_{nl}(k_+)$$

is satisfied. Thus, the group velocities of the asymptotic wave trains, computed in a frame moving with the interface, are again directed toward the interface, which illustrates further why these structures correspond to the Lax shocks of the Burgers equation.

We can only prove the existence of u_* when k_+ and k_- are close to each other[5], so that the coherent structures are really *weak* shocks. It is an interesting consequence of (1.13) that $c_* \to c_g(k_0)$ as $k_\pm \to k_0$. Thus, the speed of the interface between the two asymptotic wave trains becomes the group velocity c_g when the wave trains approach each other.

---

[5]More precisely, when both the wave numbers *and* the profiles of the associated wave trains, in the 2π-periodic θ-variable, are sufficiently close to each other.

The idea of the existence (and stability) proof is to use spatial dynamics and the Kirchgässner reduction [18, 29, 41]: If the weak shocks were independent of time, they would satisfy an ordinary differential equation (ODE) in the spatial variable $\xi = x - c_* t$. Since the weak shocks are periodic in time, a similar approach works: Writing the weak shock as $u_*(\xi, \tau) = u_*(x - c_* t, \omega_* t)$, we see that $(u, v)(\xi, \tau) = (u_*, \partial_\xi u_*)(\xi, \tau)$ has period 2π in τ and satisfies the modulated-wave equation

$$\begin{aligned} \partial_\xi u &= v \\ \partial_\xi v &= D^{-1}[\omega_* \partial_\tau u - c_* v - f(u)], \end{aligned} \tag{1.14}$$

where $(u, v)(\xi, \cdot)$ lies for each ξ in the space $H^1_{\mathrm{per}}(0, 2\pi) \times H^{1/2}_{\mathrm{per}}(0, 2\pi)$ of functions that are 2π-periodic in τ. In this spatial-dynamics formulation, wave trains correspond to periodic orbits of (1.14), while weak shocks are heteroclinic orbits that connect them. Since (1.14) is autonomous in ξ, periodic orbits of (1.14) will always have one neutral Floquet exponent. If we choose $c_* := c_g(k_0)$ to be the group velocity of the wave train with wave number k_0, then the periodic orbit corresponding to this wave train turns out to have two neutral Floquet exponents, and the vector field on the two-dimensional center manifold coincides with the equation describing viscous shock profiles in the Burgers equation. Thus, we obtain the weak shocks from the Lax shocks of the flow on the center manifold.

Next, we discuss the validity of the Burgers equation itself over time intervals $[0, T_0/\delta^2]$ for some fixed $T_0 > 0$. To illustrate the key ideas and difficulties, we focus first on the complex Ginzburg–Landau equation (CGL)

$$\partial_t A = (1 + i\alpha)\partial_{xx} A + A - (1 + i\beta) A |A|^2,$$

where $A(x, t)$ is complex-valued. For simplicity, we consider the wave train with wave number $k = 0$ which is given explicitly by

$$A_0(x, t) = e^{-i\beta t}$$

and whose group velocity vanishes. Exploiting the invariance of the CGL with respect to $A \mapsto A e^{i\gamma}$, we introduce the amplitude and phase deviations (r, ϕ) of this wave train via

$$A(x, t) = (1 + r(x, t)) e^{i(-\beta t + \phi(x, t))}.$$

Substituting this ansatz into the Ginzburg–Landau equation and using the local wave number $\psi = \partial_x \phi$ as a new variable, we obtain the equation

$$\begin{aligned} \partial_t r &= \partial_{xx} r - 2r - \psi^2 - \psi^2 r - 2\alpha(\partial_x r)\psi - \alpha \partial_x \psi - \alpha(\partial_x \psi)r - 3r^2 - r^3 \\ \partial_t \psi &= \partial_{xx} \psi + \partial_x \left(\frac{\alpha \partial_{xx} r}{1 + r} - \alpha \psi^2 + \frac{2(\partial_x r)\psi}{1 + r} - 2\beta r - \beta r^2 \right), \end{aligned} \tag{1.15}$$

which depends only on r and $\psi = \partial_x \phi$ but not on the phase ϕ itself. Thus, we have successfully decomposed the underlying PDE into an equation for the local wave number ψ and an equation for the remainder r. The equation for ψ is a conservation law, which somewhat resembles the Burgers equation, while the equation for r is exponentially damped due to the linear term $-2r$.

Given a solution $q(X, T)$ of the Burgers equation on the time interval $[0, T_0]$, we now wish to examine whether there is a solution $(r, \psi)(x, t)$ of (1.15) that stays close to the slowly-varying modulation $(0, \delta q(\delta x, \delta^2 t))$ for $t \in [0, T_0 \delta^{-2}]$ and $0 < \delta \ll 1$. Our approach follows closely the general strategy that has been developed over the past decade to establish validity of amplitude and modulation equations. First, using a formal power-series expansion in the scaled variables (X, T), we can calculate corrections to the initial approximation $(0, \delta q(\delta x, \delta^2 t))$ to obtain functions $(r_n, \psi_n)(X, T)$ that satisfy (1.15) up to residuals of the order $O(\delta^n)$ (with n chosen as large as we wish). This suggests that we try to control the full solution by making the ansatz

$$(r, \psi)(x, t) = (r_n, \psi_n)(x, t) + \delta^{n+1}(\mathcal{R}^s, \mathcal{R}^c)(x, t) \tag{1.16}$$

for solutions of (1.15), which leads to a certain evolution equation for the error $(\mathcal{R}^s, \mathcal{R}^c)$. Roughly speaking, we expect that the equation for $\mathcal{R}^s$ will again be exponentially damped, and that its solutions therefore stay bounded by Gronwall's lemma as long as its right-hand side remains bounded, while the equation for the error $\mathcal{R}^c$ will retain the conservation-law structure exhibited by (1.15). Corroborating these assertions requires diagonalizing (1.15) in an appropriate sense which we postpone until the actual proof. Instead, we shall focus on the model problem

$$\partial_t \mathcal{R}^c = \partial_{xx} \mathcal{R}^c + \partial_x \left(\delta a \mathcal{R}^c + O(\delta^2) \right), \qquad a = a(x, t) \in L^\infty(\mathbb{R} \times \mathbb{R}^+) \tag{1.17}$$

which turns out to share the key features with the actual equation for $\mathcal{R}^c$.

The issue at hands is to control the growth of the solution $\mathcal{R}^c$ of (1.17) over the time interval $[0, T_0/\delta^2]$. Using the variation-of-constant formula, we find

$$\mathcal{R}^c(t) = \int_0^t e^{\partial_{xx}(t-s)} \partial_x \left[\delta a(\cdot, s) \mathcal{R}^c(s) + O(\delta^2) \right] ds.$$

The estimate

$$\|e^{\partial_{xx} t} \partial_x\|_{L^2 \to L^2} \leq \frac{C_0}{\sqrt{t}}$$

of the linear semigroup gives

$$\begin{aligned} \|\mathcal{R}^c(t)\|_{L^2} &\leq \int_0^t \frac{C_0}{\sqrt{t-s}} \left[\delta \|a\|_{L^\infty} \|\mathcal{R}^c(s)\|_{L^2} + O(\delta^2) \right] ds \\ &\leq C_1 + \int_0^t \frac{\delta C_0}{\sqrt{t-s}} \|\mathcal{R}^c(s)\|_{L^2} ds, \qquad \forall t \in [0, T_0/\delta^2]. \end{aligned}$$

We can now use the version[6] of Gronwall's inequality proved in [22] to conclude boundedness of the error $\mathcal{R}^c(t)$. This is further illustrated by the calculation

$$\int_0^t \frac{\delta C_0}{\sqrt{t-s}} ds \leq \delta C_0 \sqrt{t} \leq C_0 \sqrt{T_0}$$

which is valid for $t \in [0, T_0/\delta^2]$.

In summary, it is the factor ∂_x in (1.17), i.e. its conservation-law structure, that allows us to conclude that the error stays bounded over the desired time interval. We shall see in §3.9 that the leading-order term in the equation for the error of the Ginzburg–Landau equation is indeed of order $O(\delta)$ so that we could not possibly infer boundedness over time intervals of length δ^{-2} if the factor ∂_x were not present.

To apply the same analysis to reaction-diffusion equations, we need to extract an equation of conservation-law form for the local wave number. This is accomplished in §5: Starting with an arbitrary slowly-varying phase function $\phi(x, t)$, we change the independent variable x via $x = y + \phi(y, t)$. The resulting PDE in the y-variable turns out to depend only on the derivatives of ϕ but not on ϕ itself. This allows us to derive an effective equation for the wave number upon using Bloch-wave transforms and mode filters.

---

[6]See Lemma 3.12 in §3.9.

The Burgers equation is formulated in terms of the wave number, and any validity result therefore needs to take into account the reconstruction of the phase Φ from the wave number $\partial_X \Phi$. This reconstruction turns out to impose a number of limitations on how well the dynamics in the reaction-diffusion system can be approximated via solutions of the associated Burgers equation: While our approximation results are uniform in the variable y that we introduced above, it turns out that we cannot expect uniform validity for $x \in \mathbb{R}$ but only for x in intervals of large but finite length. Moreover, we have to allow a global x-independent shift between the approximation and the solution. For this shift, we can only prove an $O(1)$-estimate on the time interval of the order $O(1/\delta^2)$.

We emphasize, however, that the quality of the approximation by the Burgers equation improves dramatically for solutions with additional properties. One example are solutions to the Burgers equation that converge sufficiently fast toward, possibly different, limits as $X \to \pm\infty$. In particular, this class of solutions includes sufficiently localized solutions $\Phi(X,T)$ of the integrated Burgers equation (1.9). For all these solutions, the approximation is uniform in $x \in \mathbb{R}$.

1.5. Related modulation equations

We derived the Burgers equation by choosing a scaling of the (x,t) that resembles the self-similarity scaling of the linear heat equation. Alternatively, following [23], we may also employ the ansatz

(1.18) $\quad u(x,t) = u_0(\omega_{\mathrm{nl}}(k)t - kx - \Phi(X,T)/\delta; k + \partial_X \Phi(X,T)), \quad (X,T) = (\delta x, \delta t)$

which leads to the inviscid Burgers equation[7]

(1.19) $$\partial_T q + \partial_X \omega_{\mathrm{nl}}(k+q) = 0$$

for the wave number $q = \partial_X \Phi$. Equation (1.19) is a hyperbolic conservation law and therefore allows the formation of shocks. Thus, we can only expect to obtain validity results that hold over time intervals $[0, T_1/\delta]$ where $T_1 > 0$ is sufficiently small depending on the chosen solution $q(X,T)$. This is indeed the result that we shall establish in §6 for reaction-diffusion equations. While the profile of solutions is again approximated well, its position is known only up to an error of the order $O(\|q\|^2/\delta)$. We emphasize though that this estimate is good enough to prove that the group velocity provides the speed with which perturbations are transported along the wave train. Lastly, we remark that a similar approximation result has been proved in [40] for the complex Ginzburg–Landau equation.

The description via the Burgers equation breaks down once the wave trains undergo sideband instabilities which occur when the coefficient $\lambda''_{\mathrm{lin}}(0)$ changes sign. In particular, the Burgers equation (1.8) becomes ill-posed near these instabilities, which shows the need of taking higher-order derivatives into account that regularize the equation. Depending on the next nonvanishing term in the linear dispersion relation $\lambda_{\mathrm{lin}}(\nu)$, it is either the Korteweg–de Vries equation (KdV)

$$\partial_T q - \frac{1}{6}\lambda'''_{\mathrm{lin}}(0)\partial_X^3 q + \frac{1}{2}\omega''_{\mathrm{nl}}(k)\partial_X(q^2) = 0$$

---

[7] Strictly speaking, the term inviscid Burgers equation refers to the conservation law with nonlinearity $q\partial_X q$ (i.e. with a quadratic flux). Within the context of this paper, we think of the Burgers equation as a shortcut for the modulation equation for the wave number and shall therefore, with a slight abuse of notation, always refer to (1.19) as the inviscid Burgers equation.

or the Kuramoto-Sivashinsky equation[8]

$$\partial_T q - \frac{1}{24}\lambda_{\text{lin}}''''(0)\partial_X^4 q + \kappa_2 \partial_{XX} q + \frac{1}{2}\omega_{\text{nl}}''(k)\partial_X(q^2) = 0$$

which describe the dynamics of modulated sideband-unstable wave trains. Validity results for both of these equations are presented in §7 though the results for the KdV equation are again quite unsatisfactory due to its hyperbolic nature. Since we chose not to exploit the regularity properties of the KdV equation, our results are limited to time intervals $[0, T_1 \delta^{-3}]$ where $0 < T_1 \ll 1$ is sufficiently small.

1.6. References to related works

The mathematics and physics literature contains a large body of works pertaining to phase and modulation equations. We shall put here our work in perspective by citing those papers that influenced us most but do not attempt to give a comprehensive literature review.

Howard and Kopell [23] were, to our knowledge, the first to consider phase equations in reaction-diffusion equations. They formally derived the inviscid Burgers equation for reaction-diffusion systems using multi-scale expansions similar in spirit to the approach pioneered by Whitham [56] in his work on conservative PDEs. Howard and Kopell also proved the existence of weak shocks in λ-ω systems where weak shocks satisfy an ODE. Lastly, they commented on the difficulties that arise for general reaction-diffusion systems when weak shocks are sought via the spatial-dynamics approach since the resulting equation is ill-posed.

Kuramoto [31] investigated more systematically the different types of phase equations that arise through formal multi-scale expansions depending on the symmetries of the underlying PDE and the stability and symmetry properties of the underlying periodic pattern (see also [43] for a comprehensive overview).

Other formal derivations of various phase equations for the Ginzburg–Landau equation and for general reaction-diffusion equations can be found, for instance, in [3, 4, 20, 30, 35]. For further references, we refer to the survey articles and textbooks [1, 10, 32, 34, 44, 45]. These also contain references to many of the works that have focused on multi-dimensional phase equations, starting with the seminal paper [11] by Cross and Newell.

There do not appear to be many mathematically rigorous results regarding the existence of weak shocks or the validity of phase equations. Kapitula proved in [25] the nonlinear stability, in polynomially weighted spaces, of the weak shocks in λ-ω systems that were found by Howard and Kopell. In [27], he considered the existence and stability of not necessarily weak shocks for the nearly real cubic Ginzburg–Landau equation. Van Baalen [2] proved validity of the integrated Kuramoto-Sivashinsky equation for the phase (but not the wave number) near the $k=0$ wave train of the CGL. Lastly, Melbourne and Schneider established in [39, 40] the validity of the phase diffusion equation and the inviscid Burgers equation near the $k=0$ wave train of the real and the complex Ginzburg–Landau equation, respectively.

---

[8]We shall give the precise definition of the diffusion coefficient κ_2 in §7.

CHAPTER 2

The Burgers equation

As outlined in the introduction, solutions q of the Burgers equation describe the local wave number of an underlying travelling spatially-periodic pattern. In this section, we review properties of these solutions.

2.1. Decay estimates

First, we recall the stability properties of the constant solutions $q(X,T) = q_0$ of the Burgers equation

(2.1) $$\partial_T q = \partial_{XX} q + \partial_X(q^2)$$

that were obtained in [6].

Thus, denote by $\tilde{q}(X,T)$ the deviation from the constant solution so that $q(X,T) = q_0 + \tilde{q}(X,T)$ satisfies (2.1). We see that $\tilde{q}(X,T)$ satisfies the PDE

(2.2) $$\partial_T \tilde{q} = \partial_{XX}\tilde{q} + 2q_0 \partial_X \tilde{q} + \partial_X(\tilde{q}^2)$$

which can be transformed back to (2.1) by the transformation

(2.3) $$X \mapsto X + 2q_0 T.$$

Hence, without loss of generality, we can restrict ourselves to the stability of $q = 0$ in the equation

(2.4) $$\partial_T \tilde{q} = \partial_{XX}\tilde{q} + \partial_X(\tilde{q}^2).$$

The results for general q_0 can then be obtained by transforming (2.4) into a comoving frame of reference via $X \mapsto X - 2q_0 T$.

The mean $\int_\mathbb{R} q(X,T)\,dX$ is conserved by the Burgers equation (2.4), and the subspace of functions with vanishing mean value is therefore invariant under the evolution of (2.4). Solutions to the linearized equation

$$\partial_T \tilde{q} = \partial_{XX}\tilde{q}$$

with vanishing mean satisfy the decay estimate

$$\|\tilde{q}(\cdot, T)\|_{L^\infty} \leq \frac{C}{1+T} \|\tilde{q}(\cdot, 0)\|_{L^1}$$

for some constant $C > 0$. For these decay rates, the nonlinear terms $\partial_X(\tilde{q}^2)$ turn out to be asymptotically irrelevant so that solutions to the nonlinear system (2.4) with zero mean have the same asymptotics as solutions to the linearized system with zero mean.

PROPOSITION 2.1 ([6]). *For each $\varepsilon \in (0, 1/2)$, there exist positive constants C_1, C_2 such that the following is true. If*

$$\|\tilde{q}(\cdot, 0)\|_{H^2(2)} \leq C_1, \qquad \int_{-\infty}^{\infty} \tilde{q}(X, 0)\,dX = 0,$$

then there exists an $A \in \mathbb{R}$ such that

$$\left\|(1+T)\tilde{q}\left(\sqrt{T}X,T\right) - AX\mathrm{e}^{-X^2/4}\right\|_{H^2(2)} \leq \frac{C_2}{(1+T)^{\frac{1}{2}-\varepsilon}}.$$

Consequently,

$$\|\tilde{q}(X,T)\|_{L^1} \leq \frac{C_2}{\sqrt{1+T}}, \qquad \|\tilde{q}(X,T)\|_{L^\infty} \leq \frac{C_2}{1+T}.$$

REMARK 2.2. The local phase Φ, which is related to the wave number q through $q = \partial_X \Phi$, satisfies the integrated Burgers equation

(2.5) $$\partial_T \Phi = \partial_{XX}\Phi + (\partial_X \Phi)^2.$$

For this equation, we have the following asymptotics. For each initial condition $\Phi(\cdot, 0)$ for which $\|\Phi(\cdot, 0)\|_{H^2(2)}$ is sufficiently small, there exists an $A \in \mathbb{R}$ such that

$$\left\|(1+\sqrt{T})\Phi\left(\sqrt{T}X,T\right) - A\mathrm{e}^{-X^2/4}\right\|_{H^2(2)} \leq \frac{C_2}{(1+T)^{\frac{1}{2}-\varepsilon}},$$

so that the renormalized phase converges toward a Gaussian.

Next, we consider localized solutions $\tilde{q} \in L^1$ of (2.4) with nonzero mean. The Cole–Hopf transformation

$$Q(X,T) = \mathrm{e}^{\int_{-\infty}^{X} \tilde{q}(Y,T)\,\mathrm{d}Y}$$

transforms (2.4) into the heat equation

$$\partial_T Q = \partial_{XX} Q.$$

Since the Burgers equation conserves the quantity $\int_\mathbb{R} \tilde{q}(X,T)\,\mathrm{d}X$, we are interested in the long-term profiles of solutions to the heat equation with initial conditions that satisfy

$$\lim_{X \to -\infty} Q(X,0) = 1, \qquad \lim_{X \to \infty} Q(X,0) = 1 + A.$$

The results in [6] show that

$$\lim_{T \to \infty} Q\left(\sqrt{T}X, T\right) = 1 + A\,\mathrm{erf}(X) + \mathrm{O}\left(\frac{1}{\sqrt{T}}\right).$$

Using that the inverse Cole–Hopf transformation is given by

$$\tilde{q}(X,T) = \frac{\partial_X Q(X,T)}{Q(X,T)},$$

we see that solutions $\tilde{q}$ to the Burgers equation (2.4) with localized initial conditions in $H^2(2)$ satisfy

$$\lim_{T \to \infty} \sqrt{T}\,\tilde{q}\left(\sqrt{T}X, T\right) = \frac{A\,\mathrm{erf}'(X)}{1 + A\,\mathrm{erf}(X)} = \frac{\mathrm{d}}{\mathrm{d}X}\ln(1 + A\,\mathrm{erf}(X)) =: f_A^*(X)$$

with rate $\mathrm{O}(1/\sqrt{T})$. The limiting profile $f_A^*(X)$ satisfies $\lim_{X \to \pm\infty} f_A^*(X) = 0$. Therefore, the renormalized solutions converge toward a non-Gaussian limit.

PROPOSITION 2.3 ([6]). *For $A \in \mathbb{R}$, define $f_A^*(X) := \frac{\mathrm{d}}{\mathrm{d}X}\ln(1 + A\,\mathrm{erf}(X))$. For each $\varepsilon \in (0, 1/2)$, there are positive constants C_1, C_2 such that the following is true. If $\|\tilde{q}(\cdot, 0)\|_{H^2(2)} \leq C_1$, then there exists an $A \in \mathbb{R}$ such that*

$$\left\|\sqrt{1+T}\,\tilde{q}\left(\sqrt{T}X, T\right) - f_A^*(X)\right\|_{H^2(2)} \leq \frac{C_2}{(1+T)^{1/2-\varepsilon}}.$$

Consequently,

$$\sup_{X \in \mathbb{R}} |\tilde{q}(X,T)| \leq \frac{C_2}{\sqrt{1+T}}.$$

2.2. Fronts in the Burgers equation

Next, we recall existence and stability properties of fronts of

(2.6) $$\partial_T q = \frac{1}{2}\lambda''_{\text{lin}}(0)\partial_{XX}q - \frac{1}{2}\omega''_{\text{nl}}(k)\partial_X(q^2)$$

where we assume that $\lambda''_{\text{lin}}(0) > 0$ and $\omega''_{\text{nl}}(k) \neq 0$. While an appropriate scaling of x and q would make both coefficients equal to one, we prefer to leave (2.6) as it is to make the results a little easier to apply.

Thus, we seek solutions of (2.6) of the form $q(X,T) = q_*(X - c_*T)$ where

(2.7) $$q_*(\xi) \to q_\pm, \qquad \xi \to \pm\infty$$

with q_- and q_+ being given real numbers. Upon substituting this ansatz into (2.6), we obtain

$$\frac{1}{2}\lambda''_{\text{lin}}(0)q''_* + c_*q'_* - \frac{1}{2}\omega''_{\text{nl}}(k)(q_*^2)' = 0$$

where we differentiate with respect to $\xi = X - c_*T$. Integrating in ξ, and using that $q_*(\xi) \to q_-$ as $\xi \to -\infty$ by (2.7), we get

$$\frac{1}{2}\lambda''_{\text{lin}}(0)q'_* + c_*q_* - \frac{1}{2}\omega''_{\text{nl}}(k)q_*^2 = c_*q_- - \frac{1}{2}\omega''_{\text{nl}}(k)q_-^2.$$

If we also require that $q_*(\xi) \to q_+$ as $\xi \to \infty$, we find that the wave speed c_* is necessarily given by

(2.8) $$c_* = \frac{1}{2}\omega''_{\text{nl}}(k)(q_+ + q_-),$$

so that the travelling-wave ODE becomes

(2.9) $$q'_* = \frac{\omega''_{\text{nl}}(k)}{\lambda''_{\text{lin}}(0)}(q_* - q_+)(q_* - q_-).$$

Thus, a necessary condition for obtaining a front that satisfies (2.7) is

(2.10) $$\omega''_{\text{nl}}(k)(q_+ - q_-) < 0.$$

If (2.10) is met, then the front $q_*(\xi)$ is given by

(2.11) $$q_*(\xi) = \frac{q_+ e^{-\omega''_{\text{nl}}(k)(q_+ - q_-)\xi/\lambda''_{\text{lin}}(0)} + q_-}{e^{-\omega''_{\text{nl}}(k)(q_+ - q_-)\xi/\lambda''_{\text{lin}}(0)} + 1}, \qquad \xi = X - c_*T$$

with c_* as in (2.8).

Perturbations $\tilde{q}$ of the front q_* satisfy the nonlinear equation

(2.12) $$\partial_T \tilde{q} = \frac{1}{2}\lambda''_{\text{lin}}(0)\partial_{\xi\xi}\tilde{q} + c_*\partial_\xi \tilde{q} - \frac{1}{2}\omega''_{\text{nl}}(k)\partial_\xi[2q_*\tilde{q} + \tilde{q}^2]$$

which is obtained from (2.6) by setting $q = q_* + \tilde{q}$ and transforming into the comoving frame $\xi = X - c_*T$. Initially, one may expect that the solutions $\tilde{q}$ decay only algebraically in t. However, by considering (2.12) on the space

$$\mathcal{X}_\eta = \left\{ \tilde{q};\ e^{\eta|\xi|}\tilde{q}(\xi) \in L^2(\mathbb{R}) \text{ and } e^{\eta|\xi|}\partial_\xi \tilde{q}(\xi) \in L^2(\mathbb{R}) \right\}$$

where $\eta > 0$ is sufficiently small, we obtain exponential decay rates: Indeed, Sturm–Liouville theory shows that the stationary solutions obtained in this fashion are linearly stable, with a simple eigenvalue at the origin due to translation invariance, when considered on $\mathcal{X}_\eta$ with $\eta > 0$ small. Using the spatially weighted norm on $\mathcal{X}_\eta$, spectral stability also implies nonlinear stability. We summarize these well-known facts in the following proposition (see also Figure 1.3).

PROPOSITION 2.4. *Assume that $\lambda_{\mathrm{lin}}''(0) > 0$ and $\omega_{\mathrm{nl}}''(k) \neq 0$. For any two numbers q_+ and q_- that satisfy (2.10), equation (2.6) has a unique front given by (2.11) which approaches the equilibria $q_\pm$ as $X \to \pm\infty$. The spectrum of the linearization of (2.12) about zero, considered on $\mathcal{X}_\eta$ for sufficiently small $\eta > 0$, lies in the left half-plane except for a simple eigenvalue at $\lambda = 0$. In particular, the fronts are nonlinearly stable with asymptotic phase in the spatially weighted space $\mathcal{X}_\eta$.*

In fact, there is a constant $a > 0$ such that small perturbations of the nonlinear fronts converge to zero algebraically with t^{-n} or exponentially like $\exp(-a\eta t)$ in the norms $\sup_{\xi \in \mathbb{R}} |\xi|^n |\tilde{q}(\xi)|$ and $\sup_{\xi \in \mathbb{R}} \exp(\eta|\xi|)|\tilde{q}(\xi)|$, respectively, for $0 < \eta \ll 1$ sufficiently small.

CHAPTER 3

The complex cubic Ginzburg–Landau equation

The complex cubic Ginzburg–Landau equation (CGL) in normal form is given by

(3.1) $$\partial_t A = (1 + \mathrm{i}\alpha)\partial_{xx} A + A - (1 + \mathrm{i}\beta)A|A|^2$$

where the coefficients $\alpha, \beta \in \mathbb{R}$ are real and where $x \in \mathbb{R}$, $t \geq 0$, and $A(x,t) \in \mathbb{C}$. The Ginzburg–Landau equation is a universal amplitude equation that can be derived by multiple-scales analyses: it describes slowly varying modulations, in space and time, of the amplitude of bifurcating spatially-periodic solutions in pattern-forming systems close to the threshold of their first instability.

Among the pattern-forming systems for which Ginzburg–Landau equations have been derived are reaction-diffusion equations and hydrodynamic stability problems such as the Bénard and the Taylor–Couette problem. Mathematical justifications and other aspects of the reduction to the Ginzburg–Landau equation have been investigated, for instance, in [8, 19, 37, 38, 52]. We refer to [1] and [42] for recent reviews of the physical and mathematical aspects, respectively, of the Ginzburg–Landau equation.

In this section, our goal is to prove that the dynamics of slow modulations of spectrally stable wave trains of (3.1) is approximated by the dynamics of an associated Burgers equation.

Before we embark on the analysis, we shall review related results that were obtained for the Ginzburg–Landau equation. Firstly, the nonlinear stability of spectrally stable wave trains in the sense of §2.1 has been proved in [5, 26]. The existence and stability of fronts, which connect different wave trains and are of Lax-shock type in the sense of §2.2, has been established in [5, 25] for the real CGL and in [27] for the nearly real CGL.

We are not aware of any general results on the existence and stability of weak shocks in the CGL. Such results can be obtained with the same methods that we employ in this paper: in fact, the proofs for CGL are simpler than in the general reaction-diffusion case since the existence problem reduces to an ordinary differential equation. We note, however, that the papers [13] and [27] establish the existence and stability, respectively, of Lax-type fronts for the nearly real CGL that are not necessarily weak in that the asymptotic wave numbers k_+ and k_- are not required to be close to each other.

3.1. Set-up

The complex Ginzburg–Landau equation has a family of time-periodic solutions

(3.2) $$A(x,t) = A_0(\omega_{\mathrm{nl}}(k)t - kx; k) = r(k)\mathrm{e}^{\mathrm{i}(kx - \omega_{\mathrm{nl}}(k)t)}$$

where $k, r(k), \omega_{\mathrm{nl}}(k) \in \mathbb{R}$. The amplitude r, the spatial wave number k, and the temporal frequency ω are related via

(3.3) $$r(k) = \sqrt{1-k^2}, \qquad \omega_{\mathrm{nl}}(k) = \beta + (\alpha - \beta)k^2.$$

In particular, these waves exist only for $|k| < 1$.

Spectral stability of these waves is checked as follows. Upon substituting the expression

(3.4) $$A(x,t) = A_0(\omega_{\mathrm{nl}}(k)t - kx; k) + \mathrm{e}^{\mathrm{i}(kx-\omega_{\mathrm{nl}}(k)t)} \left[a_1 \mathrm{e}^{\lambda t + \nu x} + a_2 \mathrm{e}^{\lambda t - \nu x} \right]$$

into (3.1), we see after some tedious computations that the ansatz (3.4) satisfies (3.1) to linear order in $|a_1| + |a_2|$ provided

(3.5) $$\lambda = \lambda_{\mathrm{lin}}(\nu) = -c_{\mathrm{g}}\nu + \frac{\lambda''_{\mathrm{lin}}(0)}{2}\nu^2 + \mathrm{O}(|\nu|^3)$$

with

(3.6) $$c_{\mathrm{g}} = 2k(\alpha - \beta), \qquad \frac{\lambda''_{\mathrm{lin}}(0)}{2} = 1 + \alpha\beta - \frac{2k^2(1+\beta^2)}{1-k^2}$$

(see [1, §II.D]). Therefore, if the Benjamin–Feir–Newell criterion $1 + \alpha\beta > 0$ is met, then the wave trains (3.2) are spectrally stable with respect to perturbations with small wave numbers ℓ, where $\nu = \mathrm{i}\ell$, for appropriate values of k. We remark that the wave trains are spectrally, and in fact also nonlinearly [26], stable in certain regions in (α, β, k)-space, while they are spectrally unstable with respect to finite wave numbers in other regions [20, 36].

We restrict our analysis to the following parameter regime. In particular, we do not consider the real Ginzburg–Landau equation or spectrally unstable wave trains as we shall choose $k = 0$ below.

HYPOTHESIS 3.1. *We assume $\alpha^2 + \beta^2 > 0$ and $1 + \alpha\beta > 0$.*

3.2. Slowly-varying modulations of the $k = 0$ wave train: Results

To describe slowly-varying spatio-temporal wave number modulations of the family of wave trains, we derive and validate the Burgers equation. We concentrate on modulations of the wave trains with wave number close to zero and seek solutions to (3.1) of the form

(3.7) $$\begin{aligned} A(x,t) &= A_0(\omega_{\mathrm{nl}}(0)t - \Phi(\delta x, \delta^2 t); \delta \partial_X \Phi(\delta x, \delta^2 t)) \\ &= r(\delta \partial_X \Phi(\delta x, \delta^2 t)) \mathrm{e}^{\mathrm{i}(\Phi(\delta x, \delta^2 t) - \omega_{\mathrm{nl}}(0)t)} \end{aligned}$$

where $0 < \delta \ll 1$ is a small scaling parameter. We will comment below on the differences when the basic wave number is not zero. For the above expression (3.7) to be an approximate solution of (3.1), it is then formally necessary, as we shall see in §3.3 below, that the phase Φ satisfies the phase equation

(3.8) $$\partial_T \Phi = (1 + \alpha\beta)\partial_{XX}\Phi + (\beta - \alpha)(\partial_X \Phi)^2$$

where we introduced $X = \delta x$ and $T = \delta^2 t$. Equation (3.8) can also be written as

$$2\partial_T \Phi = \lambda''_{\mathrm{lin}}(0)\partial_{XX}\Phi - \omega''_{\mathrm{nl}}(0)(\partial_X \Phi)^2$$

where ω''_{nl} is evaluated at $k = 0$.

Of course, once this phase equation has been derived on a formal level, the question that needs to be addressed is its validity. In other words, given that $\Phi(X,T)$ is a solution to (3.8), we should investigate in what sense, and over which time intervals, does (3.7) approximate a solution to the full Ginzburg–Landau equation (3.1). We will answer these questions by providing estimates of the difference of the formal approximation (3.7) and an exact solution $A(x,t)$ of the complex Ginzburg–Landau equation over time scales of the order $\mathrm{O}(1/\delta^2)$.

THEOREM 3.2. *Assume that Hypothesis 3.1 is met, and fix an integer $n \geq 3$. For each choice of $C_0 > 0$ and $T_0 > 0$ there exist constants $\delta_1 > 0$ and $C_1 > 0$ such that the following is true: For each $\delta \in (0, \delta_1)$ and each solution $\Phi(X, T)$ of (3.8) for which*

$$\text{(3.9)} \qquad \sup_{T \in [0, T_0]} \|\Phi(\cdot, T)\|_{H^n_{\mathrm{ul}}} \leq C_0,$$

there exists a solution $A = A(x,t)$ of the complex Ginzburg–Landau equation (3.1) such that

$$\sup_{t \in [0, T_0/\delta^2]} \sup_{x \in \mathbb{R}} |A(x,t) - \mathrm{e}^{\mathrm{i}[\Phi(\delta x, \delta^2 t) - \omega_{\mathrm{nl}}(0) t]}| \leq C_1 \delta^2.$$

In the preceding result, we only allow solutions Φ that satisfy (3.9). In particular, the phase function Φ is bounded in X, and we therefore do not change the wave number of the underlying wave train.

Thus, to extend the preceding result, we will now consider modulations $q = \partial_X \Phi$ of the local wave number which satisfy the Burgers equation

$$\text{(3.10)} \qquad \partial_T q = (1 + \alpha \beta) \partial_{XX} q + (\beta - \alpha) \partial_X (q^2)$$

or, equivalently,

$$2 \partial_T q = \lambda''_{\mathrm{lin}}(0) \partial_{XX} q - \omega''_{\mathrm{nl}}(0) \partial_X (q^2).$$

Of particular interest are solutions $q(X, T)$ that converge to different limits $q_\pm$ as $X \to \pm\infty$: such solutions describe the evolution of interfaces between wave trains with wave number q_- at $x = -\infty$ and q_+ at $x = \infty$. The associated phase functions Φ are, however, unbounded in X, so that Theorem 3.2 is not applicable.

THEOREM 3.3. *Assume that Hypothesis 3.1 is met, and fix integers $M \geq 3$ and $n \geq M + 3$. For each choice of $C_0 > 0$ and $T_0 > 0$, there exist constants $\delta_1 > 0$ and $C_1 > 0$ with the following property: Pick $\delta \in (0, \delta_1)$ and a solution $q(X, T)$ of the Burgers equation (3.10) for which there are numbers $q_\pm \in \mathbb{R}$ so that*

$$\sup_{T \in [0, T_0]} \left[\|q(\cdot, T)\|_{H^n_{\mathrm{ul}}} + \|(q(\cdot, T) - q_+) \rho^2_{\mathrm{w}}\|_{H^n_{\mathrm{ul}}(\mathbb{R}^+)} + \|(q(\cdot, T) - q_-) \rho^2_{\mathrm{w}}\|_{H^n_{\mathrm{ul}}(\mathbb{R}^-)} \right] \leq C_0$$

where $\rho_{\mathrm{w}}(X) = \sqrt{1 + X^2}$. Then there exists a higher-order approximation (q_h, r_h) with

$$\sup_{T \in [0, T_0]} \sup_{X \in \mathbb{R}} \left[\left| r_h(X, T) + \frac{1}{2} [q(X, T)^2 + \alpha \partial_X q(X, T)] \right| + |q_h(X, T) - q(X, T)| \right] \leq C_1 \delta$$

and a solution $A(x, t)$ of the complex Ginzburg–Landau equation (3.1) such that

$$\sup_{t \in [0, T_0/\delta^2]} \sup_{x \in \mathbb{R}} |A(x, t) - A_{\mathrm{approx}}(x, t)| \leq C_1 \delta^{M - 3/2}$$

where

$$A_{\mathrm{approx}}(x, t) = \left[1 - \frac{\delta^2}{2} r_h(\delta x, \delta^2 t)\right] \exp\left(\mathrm{i}\left[\delta q_- x + \delta \int_{-\infty}^{x} (q_h(\delta y, \delta^2 t) - q_-) \, \mathrm{d}y - \omega_{\mathrm{nl}}(0) t\right]\right).$$

We remark that the higher-order approximation (q_h, r_h) can, in principle, be computed from the solution q through the solutions of a recursive set of linear PDEs. We refer to §3.4 for details.

Somewhat surprisingly, there appear to be serious limitations regarding the quality of the approximation when the requirement that q has limits as $X \to \pm\infty$ is dropped. In particular, as we shall see in Remark 3.13, we cannot expect validity uniformly for all $x \in \mathbb{R}$ but only for x in intervals of finite length, where the length depends on the accuracy of the approximation.

THEOREM 3.4. *Assume that Hypothesis 3.1 is met. For any choice of integers $M \geq 1$ and $n \geq M + 3$, and real numbers $C_0, L_0, T_0 > $ and $0 < l < M$, there exist constants $\delta_1 > 0$ and $C_1 > 0$ such that the following is true: Pick $\delta \in (0, \delta_1)$ and a solution $q(X, T)$ of the Burgers equation (3.10) for which*

$$\sup_{T \in [0, T_0]} \|q(\cdot, T)\|_{H^n_{\mathrm{ul}}} \leq C_0,$$

then there exist a global phase function $\phi_0(t)$ with

$$\sup_{t \in [0, T_0/\delta^2]} |\phi_0(t)| \leq C_1,$$

higher-order approximations (q_h, r_h) with

$$\sup_{T \in [0, T_0]} \sup_{X \in \mathbb{R}} \left[\left| r_h(X, T) + \frac{1}{2}[q(X, T)^2 + \alpha \partial_X q(X, T)] \right| + |q_h(X, T) - q(X, T)| \right] \leq C_1 \delta,$$

and a solution $A(x, t)$ of the complex Ginzburg–Landau equation (3.1) such that

$$\sup_{t \in [0, T_0/\delta^2]} \sup_{|x| \leq L/\delta^l} |e^{-i\phi_0(t)} A(x, t) - A_{\mathrm{approx}}(x, t)| \leq C_1 \delta^{1+M-l}$$

where

$$A_{\mathrm{approx}}(x, t) = \left[1 - \frac{\delta^2}{2} r_h(\delta x, \delta^2 t)\right] \exp\left(i \left[\int_0^x \delta q_h(\delta y, \delta^2 t)\,\mathrm{d}y - \omega_{\mathrm{nl}}(0) t\right]\right).$$

The difficulty in justifying the Burgers equation for the Ginzburg–Landau equation is the time scale $O(1/\delta^2)$. Since the admissible modulations are of order $O(\delta)$, an application of Gronwall's inequality would only give validity over a time scale $O(1/\delta)$. Thus, a more refined method has to be used. We also remark that it is not obvious why an approximation result should hold for the Burgers equation (3.10): as shown in [53], there are examples of amplitude equations that are derived formally in a correct way, but that do not describe the dynamics in the original system in the desired way.

We remark that, as shown in [40], it is possible to allow modulations of order $O(1)$. In this case, the wave number modulation q satisfies the inviscid Burgers equation

$$\partial_T q + \partial_X [\omega_{\mathrm{nl}}(q)] = 0.$$

However, validity is only expected over time scales T_1/δ where, in contrast to the situation discussed in Theorem 3.4, T_1 cannot be chosen arbitrarily but comes out of the analysis: Since the inviscid Burgers equation is a conservation law, we expect the formation of shocks at which stage validity breaks down. Secondly, the approach in [40] requires analytic initial data which makes it possible to trade exponential decay in Fourier space (as $k \to \pm\infty$) for temporal decay. Lastly, the global phase difference $\phi_0(t)$ will be of order is $O(1/\delta)$ on the natural time scale of the conservation law. We shall revisit this situation in §6.

The remainder of this section is devoted to the proofs of the results stated above. We first derive the Burgers equation from the complex Ginzburg–Landau equation using polar coordinates. Afterwards, we construct higher-order approximations that are then used to formulate an approximation result for the wave numbers. The idea is now to separate the critical modes belonging to marginally stable spectral curve $\lambda_{\text{lin}}(i\ell)$ from the noncritical modes that decay in time: since the relevant eigenmodes are associated with curves of spectrum, we need to employ mode filters to accomplish this separation of modes rigorously. The key feature that we shall then exploit is that the equation derived in this fashion for the critical modes has extra structure: roughly speaking, the nonlinearity is in conservation-law form. It turns out that the same structure is present in the equations for the residuals. The next step consists in applying Gronwall's inequality to the system for the residuals: the extra conservation-law structure supplies the better estimates for the linear problem that allow us to work on the natural time scale $1/\delta^2$. Lastly, the results obtained for the wave numbers need then be transferred back to the original formulation. It is at this stage that the restrictions and limitations of Theorem 3.4 arise.

3.3. Derivation of the Burgers equation

We formally derive the Burgers equation for the complex Ginzburg–Landau equation

$$(3.11) \qquad \partial_t A = (1 + i\alpha)\partial_{xx} A + A - (1 + i\beta)A|A|^2$$

where $\alpha, \beta \in \mathbb{R}$, $x \in \mathbb{R}$, $t \geq 0$, and $A(x,t) \in \mathbb{C}$. Recall that this equation admits the wave trains

$$A(x,t) = A_0(\omega_{\text{nl}}(k)t - kx; k) = r(k)e^{i(kx - \omega_{\text{nl}}(k)t)}$$

where

$$r(k) = \sqrt{1 - k^2}, \qquad \omega_{\text{nl}}(k) = \beta + (\alpha - \beta)k^2.$$

We concentrate on long-wavelength modulations of the wave train with $k = 0$ given by

$$A(x,t) = A_0(\omega_{\text{nl}}(0)t; 0) = e^{-i\beta t},$$

and therefore introduce the amplitude and phase deviations (r, ϕ) of this wave train via

$$A(x,t) = (1 + r(x,t))e^{i(-\beta t + \phi(x,t))}.$$

The function $A(x,t)$ satisfies (3.11) if, and only if, (r, ϕ) satisfies

$$(3.12) \quad \begin{aligned} \partial_t r &= \partial_{xx} r - 2r - (\partial_x \phi)^2 - (\partial_x \phi)^2 r - 2\alpha(\partial_x r)(\partial_x \phi) \\ &\quad - \alpha\partial_{xx}\phi - \alpha(\partial_{xx}\phi)r - 3r^2 - r^3 \\ \partial_t \phi &= \partial_{xx}\phi + \alpha\frac{\partial_{xx} r}{1 + r} - \alpha(\partial_x \phi)^2 + \frac{2(\partial_x r)(\partial_x \phi)}{1 + r} - 2\beta r - \beta r^2. \end{aligned}$$

Next, we replace the equation for the phase ϕ by an equation for the local wave number $\psi = \partial_x \phi$ and obtain

$$(3.13) \quad \begin{aligned} \partial_t r &= \partial_{xx} r - 2r - \psi^2 - \psi^2 r - 2\alpha(\partial_x r)\psi - \alpha\partial_x\psi - \alpha(\partial_x \psi)r - 3r^2 - r^3 \\ \partial_t \psi &= \partial_{xx}\psi + \partial_x\left(\frac{\alpha\partial_{xx} r}{1 + r} - \alpha\psi^2 + \frac{2(\partial_x r)\psi}{1 + r} - 2\beta r - \beta r^2\right). \end{aligned}$$

We emphasize that we lose information in the process of going from phases ϕ to wave numbers $q = \partial_x \phi$ as we need to recover the value of $\phi(0,t)$. On the other hand, equation (3.13) is independent of $\phi(0,t)$.

To derive the Burgers equation, we assume that the wave number varies slowly and seek solutions of the form
$$r(x,t) = \delta^2 W(\delta x, \delta^2 t; \delta), \qquad \psi(x,t) = \delta \Psi(\delta x, \delta^2 t; \delta).$$
Substituting this ansatz into (3.13), we get

(3.14)
$$\begin{aligned}
\delta^4 \partial_T W &= \delta^2 \left[\delta^2 \partial_{XX} W - 2W - \Psi^2 - \delta^2 \Psi^2 W - 2\delta^2 \alpha (\partial_X W) \Psi \right. \\
&\quad \left. - \alpha \partial_X \Psi - \delta^2 \alpha (\partial_X \Psi) W - 3\delta^2 W^2 - \delta^4 W^3 \right] \\
\delta^3 \partial_T \Psi &= \delta^3 \left[\partial_{XX} \Psi - \partial_X (2\beta W + \alpha \Psi^2) \right. \\
&\quad \left. + \delta^2 \partial_X \left(\frac{\alpha \partial_{XX} W}{1 + \delta^2 W} + \frac{2(\partial_X W)\Psi}{1 + \delta^2 W} - \beta W^2 \right) \right]
\end{aligned}$$

where $X = \delta x$ and $T = \delta^2 t$. Dividing the two equations by δ^2 and δ^3, respectively, we get

(3.15)
$$\begin{aligned}
\delta^2 \partial_T W &= \delta^2 \partial_{XX} W - 2W - \Psi^2 - \delta^2 \Psi^2 W - 2\delta^2 \alpha (\partial_X W) \Psi \\
&\quad - \alpha \partial_X \Psi - \delta^2 \alpha (\partial_X \Psi) W - 3\delta^2 W^2 - \delta^4 W^3 \\
\partial_T \Psi &= \partial_{XX} \Psi - \partial_X (2\beta W + \alpha \Psi^2) \\
&\quad + \delta^2 \partial_X \left(\frac{\alpha \partial_{XX} W}{1 + \delta^2 W} + \frac{2(\partial_X W)\Psi}{1 + \delta^2 W} - \beta W^2 \right).
\end{aligned}$$

Neglecting terms of order $O(\delta^2)$ and higher gives the equations

(3.16)
$$\begin{aligned}
0 &= -2W_0 - \Psi_0^2 - \alpha \partial_X \Psi_0 \\
\partial_T \Psi_0 &= \partial_{XX} \Psi_0 + \partial_X (-\alpha \Psi_0^2 - 2\beta W_0)
\end{aligned}$$

for $(W_0, \Psi_0)(X,T) = (W, \Psi)(X,T;0)$, which we rewrite as
$$\begin{aligned}
W_0 &= -\frac{1}{2}(\Psi_0^2 + \alpha \partial_X \Psi_0) \\
\partial_T \Psi_0 &= (1 + \alpha\beta) \partial_{XX} \Psi_0 + (\beta - \alpha) \partial_X (\Psi_0^2).
\end{aligned}$$

Thus, every solution $q(X,T)$ of the viscous Burgers equation

(3.17)
$$\partial_T q = (1 + \alpha\beta) \partial_{XX} q + (\beta - \alpha) \partial_X (q^2)$$

gives a solution (W_0, Ψ_0) of (3.16) via

(3.18)
$$(W_0, \Psi_0) = \left(-\frac{1}{2}(q^2 + \alpha \partial_X q), q \right).$$

3.4. The construction of higher-order approximations

The higher-order approximations mentioned in Theorem 3.4 are obtained as follows. Upon writing (W, Ψ) as a formal expansion of the form

(3.19)
$$\begin{aligned}
W_M^h &= \delta^2 \left[W_0(\delta x, \delta^2 t) + \delta W_1(\delta x, \delta^2 t) + \ldots + \delta^M W_M(\delta x, \delta^2 t) \right] \\
\Psi_M^h &= \delta \left[\Psi_0(\delta x, \delta^2 t) + \delta \Psi_1(\delta x, \delta^2 t) + \ldots + \delta^M \Psi_M(\delta x, \delta^2 t) \right],
\end{aligned}$$

we find equations for (W_j, Ψ_j) which are determined from (3.15) and (3.19) as long as $\Psi_0 = q$ is given.

First, we find $\Psi_1 = W_1 = 0$. On the next level, we obtain

$$\begin{aligned}
\partial_T W_0 &= \partial_{XX} W_0 - 2W_2 - 2\Psi_0\Psi_2 - \Psi_0^2 W_0 - \alpha\partial_X\Psi_2 \\
&\quad -\alpha(\partial_X W_0)\Psi_0 - \alpha(\partial_X \Psi_0)W_0 - 3W_0^2 \\
\partial_T \Psi_2 &= \partial_{XX}\Psi_2 - \partial_X(2\beta W_2 + 2\alpha\Psi_0\Psi_2) \\
&\quad +\partial_X(\alpha\partial_{XX}W_0 + 2(\partial_X W_0)\Psi_0 - \beta W_0^2).
\end{aligned}$$

The first equation is linear in W_2 and can be solved for W_2 as a function of (W_0, Ψ_0, Ψ_2). Substituting this solution for W_2 into the second equation gives a linear inhomogeneous PDE for Ψ_2, which we can solve. As a consequence, the solutions (W_2, Ψ_2) can be computed as long as $\Psi_0 = q$ is given.

Proceeding in this fashion, we find that $\Psi_{2k+1} = W_{2k+1} = 0$ for all $k \in \mathbb{N}$. Moreover, the functions W_{2k} satisfy linear equations, while Ψ_{2k} can be found as solutions of linear inhomogeneous PDEs. These functions can be calculated as long as q is given.

We remark that had we started with a wave number $k \neq 0$ in (3.2), we would have got nonzero contributions from Ψ_{2k+1} and and W_{2k+1}.

3.5. The approximation theorem for the wave numbers

The major step in proving Theorem 3.4 is the following approximation theorem for the variables (W, Ψ).

THEOREM 3.5. *Fix integers $M \geq 1$ and $1 \leq m \leq n - 3 - M$ and choose a constant $C_0 > 0$. There are then constants $C_1 > 0$ and $\delta_1 > 0$ such that the following is true. If q is a solution of the Burgers equation (3.17) with*

$$\sup_{T \in [0, T_0]} \|q(\cdot, T)\|_{H_{ul}^n} \leq C_0,$$

then, for each $\delta \in (0, \delta_1)$, there exists a higher-order approximation $(W_M^h, \Psi_M^h) \in H_{ul}^n$, see (3.19), which satisfies

$$\sup_{T \in [0, T_0]} \|(W_M^h, \Psi_M^h)(\cdot, T) - (W_0, \Psi_0)(\cdot, T)\|_{H_{ul}^{m+1}} \leq C_1 \delta$$

with (W_0, Ψ_0) as in (3.18), and a solution (W, Ψ) of (3.15) on $[0, T_0]$ such that

$$(3.20) \quad \sup_{T \in [0, T_0]} \|(W, \Psi)(\cdot, T) - (W_M^h, \Psi_M^h)(\cdot, T)\|_{H_{ul}^{m+1} \times H_{ul}^m} \leq C_1 \delta^M.$$

Hence, we have an approximation result for the variables (W, Ψ) which is uniform in space. In particular, the limitations in the statement of Theorem 3.4 are entirely due to the reconstruction of the phase ϕ from the wave number ψ.

Sections 3.6–3.9 are devoted to the proof of Theorem 3.5. We then use Theorem 3.5 in §3.10 to prove the results stated in §3.2.

3.6. Mode filters, and separation into critical and noncritical modes

To prove Theorem 3.5, we need to separate the dynamics of the critical modes corresponding to marginally stable spectrum of the wave trains from the remaining damped modes. Upon linearizing equation (3.13) about $(r, \psi) = 0$, we obtain

$$\begin{aligned}
\partial_t r &= \partial_{xx} r - 2r - \alpha\partial_x \psi \\
\partial_t \psi &= \partial_{xx} \psi + \alpha\partial_{xxx} r - 2\beta\partial_x r
\end{aligned}$$

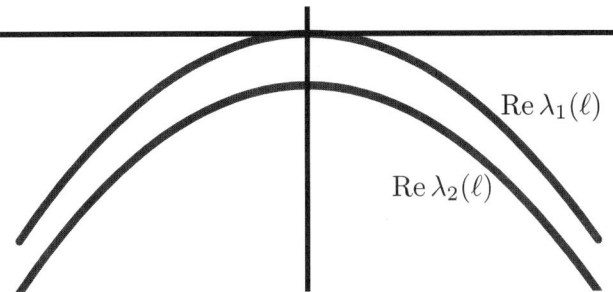

FIGURE 3.1. The real part of the eigenvalues $\lambda_j(\ell)$ is plotted as a function of the Fourier wave number ℓ for $1 + \alpha\beta > 0$.

which we consider on $H_{\mathrm{ul}}^{m+1} \times H_{\mathrm{ul}}^m$. Using the Fourier ansatz

$$\begin{pmatrix} r \\ \psi \end{pmatrix}(x,t) = \mathrm{e}^{\lambda t + \mathrm{i}\ell x} f(\ell), \qquad f(\ell) \in \mathbb{C}^2$$

with $\ell \in \mathbb{R}$, we see that f needs to satisfy the equation

$$(3.21) \qquad \Lambda(\mathrm{i}\ell) f = \lambda f, \qquad \Lambda(\mathrm{i}\ell) := \begin{pmatrix} -2 - \ell^2 & -\alpha \mathrm{i}\ell \\ -\mathrm{i}\ell(\alpha \ell^2 + 2\beta) & -\ell^2 \end{pmatrix}.$$

We obtain two spectral curves $\ell \longmapsto \lambda_j(\ell)$ where $j = 1, 2$ with $\lambda_1(0) = 0$ and $\lambda_2(0) = -2 < 0$. In particular, the curve $\lambda_1(\ell)$ coincides with the linear dispersion relation $\lambda_{\mathrm{lin}}(\mathrm{i}\ell)$ calculated in (3.5). Furthermore, we find that $\mathrm{Re}\,\lambda_j(\ell) < 0$ for $j = 1, 2$ uniformly in ℓ except, of course, for $\lambda_1(\ell)$ which takes values in a small neighbourhood of $\lambda = 0$ for $\ell \approx 0$.

We shall denote the eigenvectors $(r, \psi)(\ell)$ that belong to the rightmost curve $\lambda_1(\ell) = \lambda_{\mathrm{lin}}(\mathrm{i}\ell)$ of spectrum by $\hat{f}_1(\ell)$ and refer to them as the critical modes when $\ell \approx 0$. We also use the notation $\hat{\Lambda}(\ell)$ for the 2×2 matrix $\Lambda(\mathrm{i}\ell)$. To separate the nonlinear dynamics of the critical modes from the remaining temporally damped noncritical modes, we introduce mode filters.

First, there is an $\ell_1 > 0$ such that $\lambda_1(\ell)$ is the rightmost eigenvalue of $\hat{\Lambda}(\ell)$ for all ℓ with $|\ell| \leq \ell_1$. Thus, for all such ℓ, the integral

$$\hat{Q}^{\mathrm{c}}(\ell) = \frac{1}{2\pi \mathrm{i}} \int_\Gamma [\lambda - \hat{\Lambda}(\ell)]^{-1} \, \mathrm{d}\lambda$$

defines an $\hat{\Lambda}(\ell)$-invariant projection onto the subspace spanned by $\hat{f}_1(\ell)$, provided we choose $\Gamma \subset \mathbb{C}$ to be a small circle that surrounds $\lambda_1(\ell)$ counter-clockwise in the complex plane but does not intersect the rest of the spectrum of $\hat{\Lambda}(\ell)$. In particular, we have $\hat{Q}^{\mathrm{c}}(\ell) \hat{f}_1(\ell) = \hat{f}_1(\ell)$.

Since we wish to select only the critical modes, i.e. those belonging to $|\ell| \ll 1$, we pick a $\mathcal{C}_0^\infty$-cutoff function $\chi : \mathbb{R} \to [0, 1]$ with values in $[0, 1]$ so that

$$(3.22) \qquad \chi(\ell) = \begin{cases} 1 & \text{for } |\ell| \leq 1, \\ 0 & \text{for } |\ell| \geq 2, \end{cases}$$

and define

$$\hat{P}^{\mathrm{c}}(\ell) := \hat{Q}^{\mathrm{c}}(\ell) \chi\left(\frac{2\ell}{\ell_1}\right), \qquad \hat{P}^{\mathrm{s}}(\ell) := 1 - \hat{Q}^{\mathrm{c}}(\ell) \chi\left(\frac{8\ell}{\ell_1}\right)$$

as well as
$$\hat{P}^{\mathrm{c}}_{\mathrm{mf}}(\ell) := \hat{Q}^{\mathrm{c}}(\ell)\chi\left(\frac{4\ell}{\ell_1}\right), \qquad \hat{P}^{\mathrm{s}}_{\mathrm{mf}}(\ell) := 1 - \hat{Q}^{\mathrm{c}}(\ell)\chi\left(\frac{4\ell}{\ell_1}\right).$$

Any two of these matrices commute for each fixed ℓ, and we have

$$(3.23) \qquad (1-\hat{P}^{\mathrm{c}})\hat{P}^{\mathrm{c}}_{\mathrm{mf}} = 0, \qquad (1-\hat{P}^{\mathrm{s}})\hat{P}^{\mathrm{s}}_{\mathrm{mf}} = 0, \qquad \hat{P}^{\mathrm{c}}_{\mathrm{mf}} + \hat{P}^{\mathrm{s}}_{\mathrm{mf}} = 1$$

which shows that the operators $\hat{P}^{\mathrm{c}}_{\mathrm{mf}}$ and $\hat{P}^{\mathrm{c}}$ together behave to some extent similar to projections. Lastly, we set

$$\hat{\lambda}^{\mathrm{c}}(\ell) := \lambda_1(\ell)$$

and define scalar-valued projections $\hat{p}^{\mathrm{c}}(\ell)$ and $\hat{p}^{\mathrm{c}}_{\mathrm{mf}}(\ell)$ implicitly by

$$[\hat{p}^{\mathrm{c}}(\ell)v]\hat{f}_1(\ell) = \hat{P}^{\mathrm{c}}(\ell)v, \qquad [\hat{p}^{\mathrm{c}}_{\mathrm{mf}}(\ell)v]\hat{f}_1(\ell) = \hat{P}^{\mathrm{c}}_{\mathrm{mf}}(\ell)v, \qquad \forall v \in \mathbb{C}^2.$$

We now employ multiplier theory to transfer these operators from Fourier space back to physical space. Throughout this paper, the Fourier transform of a variable u is denoted by $\hat{u}$. To each multiplication operator $\hat{\mathcal{M}}$ in Fourier space, we associate the operator

$$(3.24) \qquad \mathcal{M}: \quad u \longmapsto \mathcal{F}^{-1}(\hat{\mathcal{M}}\mathcal{F}u)$$

where $\mathcal{F}$ denotes the Fourier transform. Thus, in x-space, we denote operators by the same symbol but with the superscript $\hat{}$ being dropped. The following multiplier theorem gives estimates for the operator $\mathcal{M}$ in the physical spaces H^m_{ul}.

LEMMA 3.6 ([52, Lemma 5]). *Let $\mathcal{W}_1, \mathcal{W}_2$ be Hilbert spaces and fix $m \in \mathbb{Z}$. If*

$$\hat{\mathcal{M}}: \quad \mathbb{R} \longrightarrow L(\mathcal{W}_1, \mathcal{W}_2), \quad \ell \longmapsto \hat{\mathcal{M}}(\ell)$$

is a map such that $\ell \mapsto (1+\ell^2)^{m/2}\hat{\mathcal{M}}(\ell)$ lies in $\mathcal{C}^2_{\mathrm{b}}(\mathbb{R}, L(\mathcal{W}_1, \mathcal{W}_2))$, then, for each $q \in \mathbb{N}_0$ with $q+m \geq 0$, the operator $\mathcal{M}$ defined through (3.24) can be extended to a bounded operator

$$\mathcal{M}: \quad H^q_{\mathrm{ul}}(\mathbb{R}, \mathcal{W}_1) \longrightarrow H^{q+m}_{\mathrm{ul}}(\mathbb{R}, \mathcal{W}_2)$$

whose norm satisfies

$$\|\mathcal{M}\|_{L(H^q_{\mathrm{ul}}(\mathbb{R},\mathcal{W}_1), H^{q+m}_{\mathrm{ul}}(\mathbb{R},\mathcal{W}_2))} \leq C(q,m)\|(1+|\cdot|^2)^{m/2}\hat{\mathcal{M}}(\cdot)\|_{\mathcal{C}^2_{\mathrm{b}}(\mathbb{R},L(\mathcal{W}_1,\mathcal{W}_2))},$$

where $C(q,m)$ does not depend on $\hat{\mathcal{M}}$.

REMARK 3.7. *The above lemma also holds for multilinear mappings.*

We record that the mappings $\hat{P}^{\mathrm{c}}_{\mathrm{mf}}$, $\hat{P}^{\mathrm{c}}$, $\hat{p}^{\mathrm{c}}_{\mathrm{mf}}$, $\hat{p}^{\mathrm{c}}$ and $\hat{f}_1$ have compact support in ℓ. Thus, an application of Lemma 3.6 with $\mathcal{W}_1 = \mathcal{W}_2 = \mathbb{C}^2$ shows that $P^{\mathrm{c}}_{\mathrm{mf}}$, P^{c}, $p^{\mathrm{c}}_{\mathrm{mf}}$, p^{c} and f_1 are bounded linear operators from $H^1_{\mathrm{ul}} \times L^2_{\mathrm{ul}}$ into $H^{m+1}_{\mathrm{ul}} \times H^m_{\mathrm{ul}}$ for each $m \geq 0$. Similarly, $P^{\mathrm{s}}_{\mathrm{mf}}$ and P^{s} are bounded linear operators on $H^{m+1}_{\mathrm{ul}} \times H^m_{\mathrm{ul}}$ for each $m \geq 0$.

We are now well prepared to establish the desired splitting into critical and noncritical modes in Fourier as well as in physical space. Using the notation $v = (r, \psi)$, we write (3.13) as

$$(3.25) \qquad \partial_t v = \Lambda v + \mathcal{N}(v).$$

We seek solutions of this equation by considering the system

$$(3.26) \qquad \begin{aligned} \partial_t \hat{w}^{\mathrm{c}}(\ell) &= \hat{\Lambda}(\ell)\hat{w}^{\mathrm{c}}(\ell) + \hat{P}^{\mathrm{c}}_{\mathrm{mf}}(\ell)\hat{\mathcal{N}}(\hat{w}^{\mathrm{c}} + \hat{w}^{\mathrm{s}})(\ell) \\ \partial_t \hat{w}^{\mathrm{s}}(\ell) &= \hat{\Lambda}(\ell)\hat{w}^{\mathrm{s}}(\ell) + \hat{P}^{\mathrm{s}}_{\mathrm{mf}}(\ell)\hat{\mathcal{N}}(\hat{w}^{\mathrm{c}} + \hat{w}^{\mathrm{s}})(\ell) \end{aligned}$$

for $(\hat{w}^c, \hat{w}^s)$, where we require that

(3.27) $\quad (1 - \hat{P}^c(\ell))\hat{w}^c(\ell) = 0, \qquad (1 - \hat{P}^s(\ell))\hat{w}^s(\ell) = 0, \qquad \forall \ell \in \mathbb{R}$

for all $t \geq 0$. Upon applying $1 - \hat{P}^c(\ell)$ and $1 - \hat{P}^s(\ell)$ to the first and second equation in (3.26), respectively, and using (3.23), we see that (3.23) leaves $\text{Fix}\,\hat{P}^c(\ell)$ and $\text{Fix}\,\hat{P}^s(\ell)$ invariant: in other words, if (3.27) holds true initially at $t = 0$, then it is met for all $t > 0$. Most importantly, given a solution $(\hat{w}^c, \hat{w}^s)$ of (3.26) which satisfies (3.27), we see from (3.23) that the function $v = w^c + w^s$ satisfies (3.25). Lastly, we record that any initial condition can indeed be decomposed into $\hat{w}^c$ and $\hat{w}^s$.

Since $\hat{w}^c(\ell)$ lies in a one-dimensional subspace of $\mathbb{C}^2$ for each fixed Fourier wave number ℓ, we may introduce the scalar function $\hat{v}^c(\ell)$ by $\hat{w}^c(\ell) = \hat{v}^c(\ell)\hat{f}_1(\ell)$. To keep the notation consistent, we also set $\hat{v}^s := \hat{w}^s$. Thus, $(\hat{v}^c, \hat{v}^s)$ satisfies

(3.28) $\quad \begin{aligned} \partial_t \hat{v}^c(\ell) &= \hat{\lambda}^c(\ell)\hat{v}^c(\ell) + \hat{p}^c_{\text{mf}}(\ell)\hat{\mathcal{N}}(\hat{v}^c \hat{f}_1 + \hat{v}^s)(\ell) \\ \partial_t \hat{v}^s(\ell) &= \hat{\Lambda}^s(\ell)\hat{v}^s(\ell) + \hat{P}^s_{\text{mf}}(\ell)\hat{\mathcal{N}}(\hat{v}^c \hat{f}_1 + \hat{v}^s)(\ell), \end{aligned}$

where $\hat{\Lambda}^s(\ell) = \hat{\Lambda}(\ell)\hat{P}^s$. By construction, we then have $p^c v^c \hat{f}_1 = v^c$ and $P^s v^s = v^s$.

Inspecting the equation for ψ in (3.13), we see that its right-hand side is of the form $\partial_x[\ldots]$ which, in Fourier space, corresponds to a term of the form $i\ell[\ldots]$. Since ψ is related to the critical modes, we may expect that this feature survives the splitting into critical and noncritical modes. The following lemma shows that this is indeed the case. As already alluded to in §3.2, we will exploit this important property in the later stages of the proofs of the approximation theorems.

LEMMA 3.8. *There exists a smooth nonlinear mapping $\mathcal{N}^c$ which maps $H^{m+1}_{\text{ul}} \times H^m_{\text{ul}}$ into H^s_{ul} for each $s \geq 0$ and a smooth function $\hat{\rho}$ with $|\hat{\rho}(\ell)| \leq C|\ell|$ for some constant $C > 0$ such that*

$$\hat{p}^c_{\text{mf}}(\ell)\hat{\mathcal{N}}(\hat{v}^c \hat{f}_1 + \hat{v}^s)(\ell) = \hat{\rho}(\ell)\hat{\mathcal{N}}^c(\hat{v}^c \hat{f}_1 + \hat{v}^s)(\ell).$$

PROOF. The eigenfunctions of

$$\Lambda(i\ell) = \begin{pmatrix} -2 & 0 \\ 0 & 0 \end{pmatrix} + \text{O}(\ell),$$

see (3.21), have the expansion

$$\hat{f}_1 = \begin{pmatrix} 0 \\ 1 \end{pmatrix} + \text{O}(\ell), \qquad \hat{f}_2 = \begin{pmatrix} 1 \\ 0 \end{pmatrix} + \text{O}(\ell)$$

near $\ell = 0$, while the adjoint eigenfunctions satisfy

$$\hat{f}_1^* = \begin{pmatrix} 0 \\ 1 \end{pmatrix} + \text{O}(\ell), \qquad \hat{f}_2^* = \begin{pmatrix} 1 \\ 0 \end{pmatrix} + \text{O}(\ell).$$

Furthermore, upon inspecting (3.15), we see that the nonlinearity $\mathcal{N}$ from (3.25) satisfies

$$\hat{\mathcal{N}} = \begin{pmatrix} \text{O}(1) \\ \text{O}(\ell) \end{pmatrix}.$$

Hence,

$$\hat{p}^c_{\text{mf}} \hat{\mathcal{N}} = \langle \hat{f}_1^*, \hat{\mathcal{N}} \rangle = \text{O}(\ell)$$

as claimed. $\square$

Using the notation of the preceding lemma, equation (3.28) becomes

$$
\begin{aligned}
(3.29)\quad \partial_t v^c &= \lambda^c v^c + \rho \mathcal{N}^c(v^c, v^s) \\
\partial_t v^s &= \Lambda^s v^s + \mathcal{N}^s(v^c, v^s),
\end{aligned}
$$

which we shall solve for (v^c, v^s) where

$$v^c \in \mathcal{X}_m^c := H_{\mathrm{ul}}^{m+1} \cap \mathrm{Rg}(p^c|_{\mathrm{Fix}\, P^c}), \qquad v^s = (r, \psi) \in \mathcal{X}_m^s := \left(H_{\mathrm{ul}}^{m+1} \times H_{\mathrm{ul}}^m\right) \cap \mathrm{Fix}\, P^s.$$

From now on, as there is little danger of confusion, we shall denote both spaces $\mathcal{X}_m^c$ and $\mathcal{X}_m^s$ simply by $\mathcal{X}_m$.

Since the variable v^c has compact support in Fourier space, it lies, in fact, in H_{ul}^s for every $s \geq 0$; more precisely, we have $v^c f_1 \in H_{\mathrm{ul}}^{s+1} \times H_{\mathrm{ul}}^s$ for each s. We also record that ρ is a possibly nonlocal linear operator that acts similar to ∂_x.

3.7. Estimates of the linear semigroups

The semigroup associated with the linear part of (3.29) has the following properties.

LEMMA 3.9. *The operators λ^c and Λ^s are sectorial in $\mathcal{X}_m$. Furthermore, there are constants $C_0, \sigma > 0$ such that the semigroups $\mathrm{e}^{\lambda^c t}$ and $\mathrm{e}^{\Lambda^s t}$ generated by these operators satisfy*

$$
\begin{aligned}
\|\mathrm{e}^{\lambda^c t}\|_{\mathcal{X}_m \to \mathcal{X}_m} &\leq C_0 \\
\|\mathrm{e}^{\lambda^c t} \rho\|_{\mathcal{X}_m \to \mathcal{X}_m} &\leq \frac{C_0}{\sqrt{t}} \\
\|\mathrm{e}^{\Lambda^s t}\|_{\mathcal{X}_m \to \mathcal{X}_m} &\leq C_0 \mathrm{e}^{-\sigma t}
\end{aligned}
$$

for all $t > 0$ and each $m \geq 0$, where ρ is the function found in Lemma 3.8 with $|\hat{\rho}(\ell)| \leq C|\ell|$.

PROOF. The operator Λ differs by a relatively bounded perturbation from the sectorial operator

$$\begin{pmatrix} \partial_{xx} & -\alpha \partial_x \\ \alpha \partial_{xxx} & \partial_{xx} \end{pmatrix}$$

and is therefore also sectorial. On account of [22], Λ generates an analytic semigroup, and the growth rates of $\mathrm{e}^{\Lambda t}$ are determined by the position of the spectrum. In particular, $\mathrm{e}^{\Lambda^s t}$ decays with strictly negative exponential rate. The factor $t^{-1/2}$ for the critical part is obtained by noticing the parabolic form of $\mathrm{Re}\,\lambda^c$ at $k = 0$ and applying Lemma 3.6 to

$$\hat{M}(\ell) = \mathrm{e}^{\delta^{-2} \hat{\lambda}^c(\delta \ell) T} \hat{\rho}(\delta \ell)$$

with $T = \delta^2 t$. □

Taking the assertions of the Lemmas 3.8 and 3.9 together, we see that we gain a factor $t^{-1/2}$ in the equation for the critical modes v^c: this factor becomes smaller for larger t, and exploiting this additional decaying factor will be crucial in the forthcoming analysis.

3.8. Estimates of the residual

We are now in a position to begin the proof of the approximation theorems. Thus, pick a solution $q(X,T)$ of the Burgers equation (3.17) which satisfies the assumptions laid out in Theorem 3.5. The issue at hands is to see whether the function $(\delta q(\delta x, \delta^2 t), 0)$ constructed from q is a good approximation of solutions (v^c, v^s) of (3.29) over sufficiently large time scales.

More generally, we may use the higher-order approximations $(\Psi_M^h, W_M^h)(X,T)$ considered in §3.4 to construct a better approximation $(\delta V^c, \delta^2 V^s)(\delta x, \delta^2 t)$ of solutions to (3.29). A measure for the quality of these approximations are the residuals $\text{Res}_c(\delta V^c, \delta^2 V^s)$ and $\text{Res}_s(\delta V^c, \delta^2 V^s)$ defined by

$$\begin{aligned}(3.30) \qquad \text{Res}_c(v^c, v^s) &:= -\partial_t v^c + \lambda^c v^c + \rho \mathcal{N}^c(v^c, v^s) \\ \text{Res}_s(v^c, v^s) &:= -\partial_t v^s + \Lambda^s v^s + \mathcal{N}^s(v^c, v^s)\end{aligned}$$

for given functions (v^c, v^s). In other words, the residuals contain all terms that do not cancel out after substituting the ansatz (v^c, v^s) into the PDE.

Equation (3.14) shows that the residuals of $(\delta q, 0)$ satisfy

$$\begin{aligned}\text{Res}_c(\delta q, 0) &= \delta^3 \left[-\partial_T q + \frac{\lambda_1''(0)}{2} \partial_{XX} q + \frac{\beta - \alpha}{2} \partial_X(q^2) \right] + O(\delta^4) = O(\delta^4) \\ \text{Res}_s(\delta q, 0) &= \mathcal{N}^s(\delta q, 0) = O(\delta^2).\end{aligned}$$

The following lemma asserts that we can always find approximations whose residuals go to zero with an arbitrarily large, but fixed, algebraic rate as δ goes to zero.

LEMMA 3.10. *Pick positive integers n, m, M with $n \geq M + m + 3$, then there exist functions $(V^c, V^s)(\delta\cdot, T) \in \mathcal{X}_n$ and positive constants $\delta_1 > 0$ and $C_{\text{res}} > 0$ such that*

$$\begin{aligned}(3.31) \qquad \sup_{T \in [0, T_0]} \|V^c(\delta\cdot, T) - q(\delta\cdot, T)\|_{\mathcal{X}_m} &\leq C_{\text{res}} \delta \\ \sup_{T \in [0, T_0]} \left(\|V^c(\delta\cdot, T)\|_{\mathcal{X}_m} + \|V^s(\delta\cdot, T)\|_{\mathcal{X}_m} \right) &\leq C_{\text{res}} \\ \sup_{t \in [0, T_0/\delta^2]} \|\text{Res}_c(\delta V^c(\delta\cdot, \delta^2 t), \delta^2 V^s(\delta\cdot, \delta^2 t))\|_{\mathcal{X}_m} &\leq C_{\text{res}} \delta^{M+3} \\ \sup_{t \in [0, T_0/\delta^2]} \|\text{Res}_s(\delta V^c(\delta\cdot, \delta^2 t), \delta^2 V^s(\delta\cdot, \delta^2 t))\|_{\mathcal{X}_m} &\leq C_{\text{res}} \delta^{M+2}\end{aligned}$$

is true for all $\delta \in (0, \delta_1)$.

PROOF. Similar to the calculations in §3.4, we expand (v^c, v^s) into a formal series of the form

$$\begin{aligned}\delta V^c &= \delta(V_0^c + \delta V_1^c + \ldots + \delta^M V_M^c) \\ \delta^2 V^s &= \delta^2(V_0^s + \delta V_1^s + \ldots + \delta^M V_M^s)\end{aligned}$$

and find equations for $(V_j^c, V_j^s)(\delta x, \delta^2 t)$, which we subsequently solve recursively for given $V_0^c = q$ and $V_0^s = 0$. In detail, we use the relations

$$\partial_t V_j^c = \delta^2 \partial_T V_j^c$$
$$\partial_t V_j^s = \delta^2 \partial_T V_j^s$$
$$\lambda^c V_j^c = \frac{\delta^2 \partial_{\nu\nu}\hat{\lambda}^c(0)}{2}\partial_{XX}V_j^c + O(\delta^3)$$
$$\Lambda V_j^s = \hat{\Lambda}(0)V_j^s + O(\delta)$$
$$\rho \mathcal{N}^c = \delta \hat{\rho}'(0)\partial_X \mathcal{N}^c + O(\delta^3)$$

and obtain the system

$$\partial_T V_0^c = \frac{\partial_{\nu\nu}\hat{\lambda}^c(0)}{2}\partial_{XX}V_0^c - \frac{1}{2}(\alpha - \beta)\partial_X(V_0^c)^2$$
$$0 = \hat{\Lambda}^s(0)V_0^s + O(\|V_0^c\|^2)$$

for V_0^c and V_0^s. Since $\Lambda^s(0) = -2$, we can solve the second equation for V_0^s, while the first equation is the Burgers equation as claimed.

Next, the equations for V_1^c and V_1^s are of the form

$$\partial_T V_1^c = \frac{\partial_{\nu\nu}\hat{\lambda}^c(0)}{2}\partial_{XX}V_1^c + O(|V_0^c|(1 + |V_1^c| + |V_0^s| + |V_0^c|))$$
$$0 = \hat{\Lambda}^s(0)V_1^s + O(|V_0^c|(1 + |V_1^c| + |V_0^s| + |V_0^c|)).$$

Again, the second equation can be solved for V_1^s. The first equation is a linear parabolic PDE in V_1^c since the higher-order terms contain at most first-order derivatives of V_1^c. Thus, V_1^s and V_1^c exist as long as V_0^c is given.

A similar analysis can be carried out recursively for all V_j^s and V_j^c. As a consequence, the residual can be made as small as we wish with respect to powers of δ. The assertions about regularity follow by counting derivatives in the preceding expansions of λ^c, Λ^s and the nonlinearities. □

We remark that, from now on, we do not distinguish in our notation whether the approximations V^c and V^s are regarded as functions in (X, T) or (x, t) through $(X, T) = (\delta x, \delta^2 t)$.

3.9. Estimates of the errors

Having constructed initial approximations with very small residuals, it remains to prove that the error, defined as the difference between genuine solutions of (3.29) and the initial approximations, is also small. Anticipating the expected scaling of the errors, we define the critical and noncritical parts R^c and R^s of the error via

$$v^c(x, t) = \delta V^c(\delta x, \delta^2 t) + \delta^{M+1} R^c(x, t)$$
$$v^s(x, t) = \delta^2 V^s(\delta x, \delta^2 t) + \delta^{M+2} R^s(x, t),$$

where we assume that the left-hand sides satisfy (3.29). Substituting this ansatz into (3.28), we obtain the system

(3.32)
$$\partial_t R^c = \lambda^c R^c + \rho g^c(R^c, R^s, t)$$
$$\partial_t R^s = \Lambda^s R^s + g^s(R^c, R^s, t)$$

for (R^c, R^s). We remark that we shall always work with the initial data
$$(R^c, R^s)|_{t=0} = 0. \tag{3.33}$$
We claim that the nonlinear terms satisfy the estimates
$$\begin{aligned}
\|g^c(R^c, R^s, t)\|_{\mathcal{X}_m} &\leq \delta C_1 \|R^c\|_{\mathcal{X}_m} + \delta^2 C_{\text{Res}} \\
&\quad + \delta^2 C_1 \|R^s\|_{\mathcal{X}_m} + \delta^{M+1} C_2(D_c, D_s) \\
\|g^s(R^c, R^s, t)\|_{\mathcal{X}_{m-2}} &\leq C_{\text{Res}} + C_1 \|R^c\|_{\mathcal{X}_m} \\
&\quad + \delta C_1 \|R^s\|_{\mathcal{X}_m} + \delta^M C_2(D_c, D_s)
\end{aligned} \tag{3.34}$$
for
$$\|R^c\|_{\mathcal{X}_m} \leq D_c, \qquad \|R^s\|_{\mathcal{X}_m} \leq D_s \tag{3.35}$$
with constants D_c and D_s that are independent of $0 < \delta \ll 1$ and will be chosen later. The specific form of (3.34) arises as follows: First, quadratic interactions of δV^c with $\delta^{M+1} R^c$ lead to $O(\delta)$ terms[1] in the equation for R^c and to $O(1)$ terms in the equation for R^s. Next, quadratic interactions with $\delta^2 V^s$ lead again to higher-order terms, while quadratic interactions of $\delta^{M+1} R_c$ with itself leads to $O(\delta^{M+1})$ terms in the equation for R^c and to $O(\delta^M)$ terms in the equation for R^s. Lemma 3.8 shows that we do not have any loss of regularity for the nonlinearity g^c.

Our goal is to prove that the solution (R^c, R^s) of (3.32)–(3.33) stays bounded, uniformly in $0 < \delta \ll 1$, over the time interval $[0, T_0/\delta^2]$.

Since the nonlinearity g^s is bounded on bounded sets, and given the exponential decay of the semigroup generated by the linear operator Λ^s, we do not expect any problems from the second equation in (3.32). However, Lemma 3.9 shows that the semigroup associated with the critical part λ^c is merely bounded: thus, the term $C_1 \delta \|R^c\|_{\mathcal{X}_m}$ in the estimate (3.34) of the nonlinearity g^c may cause trouble when integrated over $[0, T_0/\delta^2]$. It is here where the special structure of (3.32) becomes crucial. If we subsume the factor ρ in (3.32) into the semigroup generated by λ^c, we can exploit the improved estimate
$$\|e^{\lambda^c t} \rho\|_{\mathcal{X}_m \to \mathcal{X}_m} \leq C_0 t^{-1/2}$$
which we established in Lemma 3.9. Integrating $t^{-1/2}$ over $[0, T_0/\delta^2]$ gives only a contribution of order $1/\delta$, instead of the factor $1/\delta^2$ when integrating a constant, which can now be taken care of by the term $C_1 \delta \|R^c\|_{\mathcal{X}_m}$.

In the remaining part of this section, we carry out the details. We begin by recording that (3.13) and (3.29), and therefore also (3.32), are quasilinear. This is further reflected in the estimate (3.34) which shows that the nonlinear terms map $\mathcal{X}_m$ into $\mathcal{X}_{m-2}$ (and not into $\mathcal{X}_{m-1}$). Thus, we shall treat (3.32) as a fully nonlinear equation making extensive use of the results in [33] regarding long-time existence, uniqueness and optimal regularity of solutions. First, we have the following estimate.

LEMMA 3.11 ([33, Theorems 4.3.1(iii) & 4.4.1(ii)]). *Fix $0 < \gamma < 1$, then there exists a constant $C_3 > 0$ with the following property: For each t_1 with $0 < t_1 \leq \infty$ and each function $\mathcal{N}^s \in \mathcal{C}^{0,\gamma}([0, t_1], \mathcal{X}_{m-2})$ with $\mathcal{N}^s(t) = P^s \mathcal{N}^s(t)$ and $\mathcal{N}^s(0) \in \mathcal{X}_m$, there is a unique solution R^s of*
$$\partial_t R^s = \Lambda^s R^s + \mathcal{N}^s(t), \qquad R^s|_{t=0} = 0$$

---

[1] The term $-\alpha \partial_X(\psi^2)$ in the second equation in (3.15) shows that this quadratic term will indeed be present.

on $[0, t_1]$, and
$$\|R^s\|_{\mathcal{C}^{0,\gamma}([0,t_1],\mathcal{X}_m)} \leq C_3 \|\mathcal{N}^s\|_{\mathcal{C}^{0,\gamma}([0,t_1],\mathcal{X}_{m-2})}.$$

Since R^c has compact support in Fourier space and the operator λ^c in the first equation in (3.32) is bounded, there is a constant $C_5 > 0$ such that

(3.36) $\quad \|R^c\|_{\mathcal{C}^{0,\gamma}([0,t_1],\mathcal{X}_m)} \leq C_5 \|R^c\|_{\mathcal{C}^{0}([0,t_1],\mathcal{X}_m)} + \|\mathcal{N}^c\|_{\mathcal{C}^{0}([0,t_1],\mathcal{X}_m)}$

for the solution R^c of
$$\partial_t R^c = \lambda^c R^c + \mathcal{N}^c(t), \qquad R^c|_{t=0} = 0$$
on $[0, t_1]$ where $\mathcal{N}^c(t) = P^c \mathcal{N}^c(t)$. The constant C_5 does not depend on t_1.

The results in [33, §8.1], and specifically [33, Theorems 8.1.1 & 8.1.3, Proposition 8.2.1], show that the nonlinear system (3.32)–(3.33) can be solved as long as the $\mathcal{C}^{0,\gamma}([0,t],\mathcal{X}_m)$-norm of the solution (R^c, R^s) stays bounded: To apply the cited results, we need to check the compatibility condition $g^s(0,0,0) \in \mathcal{X}_m$ (see also Lemma 3.11) which is met since the residuals are sufficiently smooth due to Lemma 3.10. In the rest of the proof, we shall also exploit that the Hölder norm of the residuals is bounded uniformly in δ.

We define
$$S^c(t) = \|R^c\|_{\mathcal{C}^{0,\gamma}([0,t],\mathcal{X}_m)}, \qquad S^s(t) = \|R^s\|_{\mathcal{C}^{0,\gamma}([0,t],\mathcal{X}_m)},$$
and record that these functions increase monotonically in t. We claim that $S^c(t)$ and $S^s(t)$ stay bounded uniformly in t for $t \leq T_0/\delta^2$ and $\delta \in (0, \delta_1)$ for some fixed $\delta_1 > 0$. To prove the claim, we argue by contradiction. Thus, we shall pick appropriate positive constants S_*^c and S_*^s, which will be chosen in (3.40) below independently of δ, and assume that there are sequences $\delta_j \searrow 0$ and $t_j < T_0/\delta_j^2$ so that $S^c(t) < S_*^c$ and $S^s(t) < S_*^s$ for all $0 < t < t_j$ and $S^c(t_j) \geq S_*^c$ and $S^s(t_j) \geq S_*^s$.

We proceed as follows to reach the desired contradiction. Throughout the forthcoming estimates, we restrict ourselves to $0 \leq t \leq t_j \leq T_0/\delta_j^2$. Applying the variation-of-constant formula to the equation for R^c in (3.32)–(3.33), we get
$$R^c(t) = \int_0^t e^{\lambda^c(t-\tau)} \rho g^c(R^c, R^s, \cdot)(\tau) \, d\tau.$$

The estimates provided in (3.34) and Lemma 3.9 show that
$$\begin{aligned}
\|R^c(t)\|_{\mathcal{X}_m} &\leq \int_0^t \|e^{\lambda^c(t-\tau)} \rho\|_{L(\mathcal{X}_m, \mathcal{X}_m)} \|g^c(R^c, R^s, \cdot)(\tau)\|_{\mathcal{X}_m} \, d\tau \\
&\leq \int_0^t C_0 (t-\tau)^{-1/2} \left[\delta C_1 \|R^c(\tau)\|_{\mathcal{X}_m} + \delta^2 C_{\text{Res}} \right. \\
&\quad \left. + \delta^2 C_1 \|R^s(\tau)\|_{\mathcal{X}_m} + \delta^{M+1} C_2(S_*^c, S_*^s) \right] d\tau \\
&\leq \sqrt{T_0} C_0 \left[\delta C_{\text{Res}} + \delta C_1 S_*^s + \delta^M C_2(S_*^c, S_*^s) \right] \\
&\quad + \int_0^t \frac{\delta C_0 C_1}{\sqrt{t-\tau}} \|R^c(\tau)\|_{\mathcal{X}_m} \, d\tau.
\end{aligned}$$

Using the rescaled time variable $T := \delta^2 t$, we see that

(3.37) $\quad \|R^c(T/\delta^2)\|_{\mathcal{X}_m} \leq \sqrt{T_0} C_0 \left[\delta C_{\text{Res}} + \delta C_1 S_*^s + \delta^M C_2(S_*^c, S_*^s) \right]$
$$+ \int_0^T \frac{C_0 C_1}{\sqrt{T-\tau}} \|R^c(\tau/\delta^2)\|_{\mathcal{X}_m} \, d\tau.$$

for $T \leq T_j := t_j \delta_j^2 \leq T_0$. We are now ready to apply the following version of Gronwall's inequality.

LEMMA 3.12 ([22, Lemma 7.1.1]). *Assume that b and T_0 are positive constants and $0 < d < 1$, then there is a constant $C_4 = C(b, d, T_0)$ such that the following is true. Suppose that $a \geq 0$ and $u : [0, T_0] \to \mathbb{R}^+$ is continuous with*

$$u(T) \leq a + \int_0^T \frac{bu(\tau)}{(T-\tau)^d}\, d\tau$$

for all $T \in [0, T_0]$, then

$$\sup_{T \in [0,T_0]} u(T) \leq a C_4.$$

Applying Lemma 3.12 to (3.37) with $u(t) := \|R^c(t/\delta^2)\|_{\mathcal{X}_m}$, we get

$$\|R^c(t)\|_{\mathcal{X}_m} \leq \sqrt{T_0} C_0 C_4 \left[\delta C_{\text{Res}} + \delta C_1 S_*^s + \delta^M C_2(S_*^c, S_*^s)\right].$$

Using (3.36), we finally get

(3.38) $$S^c(t) \leq \sqrt{T_0} C_0 C_4 C_5 \left[\delta C_{\text{Res}} + \delta C_1 S_*^s + \delta^M C_2(S_*^c, S_*^s)\right]$$

for all t with $0 \leq t \leq t_j$.

Next, we focus on the equation for R^s in (3.32)–(3.33). Applying Lemma 3.11 together with the estimates from (3.34) and Lemma 3.9, we obtain

$$\begin{aligned}
S^s(t) &\leq C_3 \|g^s(R^c, R^s, \cdot)\|_{\mathcal{C}^{0,\gamma}([0,t],\mathcal{X}_{m-2})} \\
&\leq C_3 \left(C_{\text{Res}} + S^c(t) + \delta S^s(t) + \delta^M C_2(S_*^c, S_*^s)\right) \\
(3.39) \quad &\leq C_3 \left(C_{\text{Res}} + S_*^c + \delta S_*^s + \delta^M C_2(S_*^c, S_*^s)\right)
\end{aligned}$$

for all t with $0 \leq t \leq t_j$.

In summary, if we choose

(3.40) $$S_*^c = 2, \qquad S_*^s = 2 + C_3(2 + C_{\text{Res}}),$$

then we get from (3.38) and (3.39) together with the assumption $\delta_j \to 0$ that

$$S^c(t_j) \leq 1, \qquad S^s(t_j) \leq 1 + C_3(2 + C_{\text{Res}})$$

for all sufficiently large j. This is the desired contradiction which establishes that the solutions (R^c, R^s) of (3.32)–(3.33) exist on $[0, T_0/\delta^2]$ and are bounded uniformly in δ.

3.10. Proofs of the theorems from §3.2

3.10.1. Proof of Theorem 3.4.
We use Theorem 3.5 to prove Theorem 3.4. The starting point of our analysis is the relation

$$A(x,t) = (1 + \delta^2 W(\delta x, \delta^2 t)) \exp\left(i\delta \int_0^x \Psi(\delta y, \delta^2 t)\, dy + i\phi_0(t) - i\beta t\right)$$

which defines a solution A of the complex Ginzburg–Landau equation (3.11) in terms of an arbitrary solution (W, Ψ) of (3.15). We wish to compare this solution with the function

$$A_{\text{approx}}(x,t) = (1 + \delta^2 W_M^{\text{h}}(\delta x, \delta^2 t)) \exp\left(i\delta \int_0^x \Psi_M^{\text{h}}(\delta y, \delta^2 t)\, dy - i\beta t\right)$$

where (W_M^h, Ψ_M^h) are the improved approximations obtained from a solution q of the Burgers equation (3.17) via the expressions (W_0, Ψ_0) from (3.18). We estimate the difference of A and A_{approx} as follows:

$$|e^{-i\phi_0(t)} A(x,t) - A_{\text{approx}}(x,t)|$$

$$\leq \left| (1+\delta^2 W(\delta x, \delta^2 t)) \exp\left(i\delta \int_0^x \Psi(\delta y, \delta^2 t)\, dy - i\beta t\right) \right.$$
$$\left. - (1+\delta^2 W_M^h(\delta x, \delta^2 t)) \exp\left(i\delta \int_0^x \Psi_M^h(\delta y, \delta^2 t)\, dy - i\beta t\right) \right|$$

$$\leq \left| (1+\delta^2 W(\delta x, \delta^2 t)) \exp\left(i\delta \int_0^x \Psi(\delta y, \delta^2 t)\, dy\right) \right.$$
$$\left. - (1+\delta^2 W(\delta x, \delta^2 t)) \exp\left(i\delta \int_0^x \Psi_M^h(\delta y, \delta^2 t)\, dy\right) \right|$$
$$+ \left| (1+\delta^2 W(\delta x, \delta^2 t)) \exp\left(i\delta \int_0^x \Psi_M^h(\delta y, \delta^2 t)\, dy\right) \right.$$
$$\left. - (1+\delta^2 W_M^h(\delta x, \delta^2 t)) \exp\left(i\delta \int_0^x \Psi_M^h(\delta y, \delta^2 t)\, dy\right) \right|$$

$$\leq \left| 1 + \delta^2 W(\delta x, \delta^2 t) \right| \times$$
$$\times \left| \exp\left(i\delta \int_0^x \Psi(\delta y, \delta^2 t)\, dy\right) - \exp\left(i\delta \int_0^x \Psi_M^h(\delta y, \delta^2 t)\, dy\right) \right|$$
$$+ \delta^2 \left| W(\delta x, \delta^2 t) - W_M^h(\delta x, \delta^2 t) \right|$$

$$\overset{\text{Thm. 3.5}}{\leq} C\delta \left| \int_0^x |\Psi(\delta y, \delta^2 t) - \Psi_M^h(\delta y, \delta^2 t)|\, dy \right| + C\delta^{M+2}$$

$$\overset{\text{Thm. 3.5}}{\leq} \left| \int_0^x C\delta^{M+1}\, dy \right| + C\delta^{M+2}$$

$$\leq C\delta(\delta^{M+1} + \delta^M |x|).$$

Restricting x to the region $|x| \leq L\delta^{-l}$, we obtain the desired estimate of the left-hand side by $C\delta^{M-l}$. The limitation comes from the phase which is obtained as the integral of the wave number which is not necessarily integrable over $\mathbb{R}$.

Lastly, inspecting the right-hand side of the ϕ-equation in (3.12), given by

$$\partial_t \phi = \partial_{xx}\phi + \alpha \frac{\partial_{xx} r}{1+r} - \alpha(\partial_x \phi)^2 + \frac{2(\partial_x r)(\partial_x \phi)}{1+r} - 2\beta r - \beta r^2,$$

we see that $\partial_t \phi(0,t) = O(\delta^2)$ and therefore

$$\sup_{t \in [0, T_0/\delta^2]} |\phi(0,t)| = O(1).$$

REMARK 3.13. We emphasize that, in general, we cannot expect better approximation properties. To see this, take $\Psi = \delta^2$ and $q = 0$. In the above estimates, we would then need to prove that the difference of $e^{i\delta^2 x}$ and 1 is smaller than $o(\delta)$. A uniform estimate over $\mathbb{R}$ can only be expected for special solutions.

3.10.2. Proof of Theorem 3.3. The idea of the proof is as follows. By assumption, the solution $q(X,T)$ of the Burgers equation that we start with converges to constants algebraically as $x \to \pm\infty$, and our ansatz therefore satisfies the complex Ginzburg–Landau equation at $x = \pm\infty$. We shall see that this implies that the residuals are also algebraically localized, which enables us to solve the equation for the errors in appropriate spaces of algebraically localized functions. When we then transfer the approximation result from wave numbers to phases, we can exploit the algebraic $1/x^2$-decay of the wave number to see that the phase, computed as the integral of the wave number in x over $\mathbb{R}$, stays bounded.

Thus, we introduce the space $H^m(n;\delta)$ which is equal to $H^m(n)$ but equipped with the norm
$$\|u\|_{H^m(n;\delta)} := \|u(\cdot)\rho_{\mathrm{w}}^n(\delta\cdot)\|_{H^m}$$
where $\rho_{\mathrm{w}}(x) = \sqrt{1+x^2}$. To prove Theorem 3.3, we first repeat the steps in the proof of Theorem 3.5. It is not hard to see that the residuals lie in $\tilde{\mathcal{X}}_m := H^{m+1}(2;\delta) \times H^m(2;\delta)$. Furthermore, the estimates in Lemma 3.10 stay the same except that the terms δ^{M+3} and δ^{M+2} on the right-hand side of the last two equations in (3.31) are replaced by $\delta^{M+5/2}$ and $\delta^{M+3/2}$, respectively, due to the scaling properties of L^2-spaces. Next, Lemma 3.11 and the estimates (3.34) of the nonlinear terms remain true in $\tilde{\mathcal{X}}_m$. Likewise, the estimates of the semigroups in Lemma 3.9 are true in $\tilde{\mathcal{X}}_m$ since the δ-dependent norm in $H^m(n;\delta)$ ensures that the constants arising in the estimates of the critical semigroup remain are $\mathrm{O}(1)$ in δ over the long time scale $\mathrm{O}(1/\delta^2)$. Thus, we see that Theorem 3.5 remains true with $\mathcal{X}_m$ replaced by $\tilde{\mathcal{X}}_m$ except that the term δ^M on the right-hand side of (3.20) needs to be replaced by $\delta^{M-1/2}$.

It remains to transfer the result from wave numbers to phases. Without loss of generality, we may assume that $q_- = 0$. We now have to compare the solution
$$A(x,t) = (1 + \delta^2 W(\delta x, \delta^2 t)) \exp\left(\mathrm{i}\delta \int_{-\infty}^x \Psi(\delta y, \delta^2 t)\,\mathrm{d}y - \mathrm{i}\beta t\right)$$
of the complex Ginzburg–Landau equation (3.11) with the approximation
$$A_{\mathrm{approx}}(x,t) = (1 + \delta^2 W_M^{\mathrm{h}}(\delta x, \delta^2 t)) \exp\left(\mathrm{i}\delta \int_{-\infty}^x \Psi_M^{\mathrm{h}}(\delta y, \delta^2 t)\,\mathrm{d}y - \mathrm{i}\beta t\right).$$
The key is that the integrals in the expressions above exist, and are $\mathrm{O}(1)$ in δ, since
$$\left|\delta \int_{-\infty}^x (1+\delta^2 y^2)^{-1}\,\mathrm{d}y\right| \leq \int_{-\infty}^\infty (1+y^2)^{-1}\,\mathrm{d}y < \infty.$$
Thus, as in §3.10.1, we obtain
$$\begin{aligned}
|A(x,t) - A_{\mathrm{approx}}(x,t)| &\leq C\delta \left|\int_{-\infty}^x [\Psi(\delta y, \delta^2 t) - \Psi_M^h(\delta y, \delta^2 t)]\,\mathrm{d}y\right| + C\delta^{M+2} \\
&\leq C\delta \int_{-\infty}^x |\delta^{M-1/2}(1+(\delta y)^2)^{-2}|\,\mathrm{d}y + C\delta^{M+2} \\
&\leq C\delta^{M-3/2}
\end{aligned}$$
uniformly for $x \in \mathbb{R}$.

3.10.3. Proof of Theorem 3.2. To prove Theorem 3.2, we follow the same strategy as in the proof of Theorem 3.5. However, instead of using equation (3.13) for the wave number ψ as in §3.5, we use only the equation (3.12) for the phase ϕ. This is feasible, of course, since we begin with a solution $\Phi(X,T)$ of the phase equation (3.8) rather than with a solution $q(X,T)$ of the Burgers equation (3.10). Thus, we shall focus on (3.12) and go through the analysis presented in §3.5 to work out the adjustments that we need to make.

3.10. PROOFS OF THE THEOREMS FROM §3.2

The major difference is that Lemma 3.8 is no longer true for (3.12), since the leading order term in the critical part of the nonlinearity is now given by $(\partial_x v^c)^2$, see (3.8). In particular, we can no longer use the improved estimate in terms of $t^{-1/2}$ for the critical part of the semigroup in Lemma 3.9 but only the bound by a constant. Thus, for the estimates of the errors in §3.9 to go through, we shall need that the nonlinearity itself is of order $O(\delta^2)$: the key is that the term $(\partial_x v^c)^2$, is indeed of the desired order $O(\delta^2)$ if v^c is a function of δx. Therefore, to exploit this property, we introduce the space $\mathcal{X}_{1,\delta}$ which, as a vector space, is equal to $\mathcal{X}_1$ but whose norm is defined via

$$\|u(\cdot)\|_{\mathcal{X}_{1,\delta}} := \|u(\delta^{-1}\cdot)\|_{\mathcal{X}_1}.$$

In this space, we obtain the estimate

$$\|e^{\lambda^c t}\|_{\mathcal{X}_{0,\delta} \to \mathcal{X}_{1,\delta}} \leq C\left(1 + \frac{1}{\delta\sqrt{t}}\right) \tag{3.41}$$

for the critical part of the linear semigroup. As in §3.9, we make the ansatz

$$\begin{aligned} v^c &= V^c + \delta^M R^c \\ v^s &= \delta^2 V^s + \delta^{M+2} R^s \end{aligned} \tag{3.42}$$

and seek functions R^c and R^s in $\mathcal{X}_{1,\delta}$ and $\mathcal{X}_m$, respectively. Note that since R^c has compact support in Fourier space, we have that $R^c \in \mathcal{X}_m$. Furthermore, the approximations V^c and V^s lie in both $\mathcal{X}_{1,\delta}$ and $\mathcal{X}_m$. Upon substitution our ansatz into the analogue of (3.28), we obtain the equation

$$\begin{aligned} \partial_t R^c &= \lambda^c R^c + g^c(R^c, R^s, t) \\ \partial_t R^s &= \Lambda^s R^s + g^s(R^c, R^s, t) \end{aligned}$$

for the errors, where the nonlinearity obeys the estimate

$$\begin{aligned} \|g^c(R^c, R^s, t)\|_{\mathcal{X}_{0,\delta}} &\leq \delta^2 C_{\text{Res}} + \delta^2 C\|R^c\|_{\mathcal{X}_{1,\delta}} + \delta^3 C\|R^s\|_{\mathcal{X}_m} \\ &\quad + \delta^M C(D_c, D_s)[\|R^c\|_{\mathcal{X}_{1,\delta}} + \|R^s\|_{\mathcal{X}_m}]^2 \\ \|g^s(R^c, R^s, t)\|_{\mathcal{X}_{m-2}} &\leq C_{\text{Res}} + C\|R^c\|_{\mathcal{X}_{1,\delta}} + \delta C\|R^s\|_{\mathcal{X}_m} \\ &\quad + \delta^M C(D_c, D_s)[\|R^c\|_{\mathcal{X}_{1,\delta}} + \|R^s\|_{\mathcal{X}_m}]^2 \end{aligned}$$

when restricted to

$$\|R^c\|_{\mathcal{X}_{1,\delta}} \leq D_c \quad \text{and} \quad \|R^s\|_{\mathcal{X}_m} \leq D_s$$

with constants D_c and D_s as in §3.9. We record that we used here that the right-hand side depends only on $\partial_x V^c = O(\delta)$ but not on $V^c = O(1)$, see (3.12).

From this point on, the rest of the proof follows exactly as in §3.9 upon utilizing the estimate (3.41) instead of the estimates provided in Lemma 3.9.

CHAPTER 4

Reaction-diffusion equations: Set-up and results

4.1. The abstract set-up

Consider the reaction-diffusion system

(4.1) $$\partial_t u = D\partial_{xx} u + f(u)$$

where $u \in \mathbb{R}^d$, $x \in \mathbb{R}$, D is a diagonal matrix with strictly positive entries, and $f : \mathbb{R}^d \to \mathbb{R}^d$ is smooth.

We assume that, for some nonzero temporal frequency $\omega = \omega_0$ and a certain nonzero spatial wave number $k = k_0$, there exists a nonconstant travelling wave train $u(x,t) = u_0(\omega_0 t - k_0 x)$ of (4.1) where $u_0(\theta)$ is 2π-periodic in its argument. Substituting this ansatz into (4.1), we see that $u_0(\theta)$ must be a 2π-periodic solution of the ordinary differential equation

(4.2) $$k^2 D\partial_{\theta\theta} u - \omega \partial_\theta u + f(u) = 0$$

for $\omega = \omega_0$ and $k = k_0$. Linearizing this equation about u_0, we obtain the linear operator $\mathcal{L}_0$, given by

(4.3) $$\mathcal{L}_0 = k^2 D\partial_{\theta\theta} - \omega \partial_\theta + f'(u_0(\theta)),$$

again with $\omega = \omega_0$ and $k = k_0$, which defines a closed and densely defined operator on $L^2_{\text{per}}(0, 2\pi)$ with domain $\mathcal{D}(\mathcal{L}_0) = H^2_{\text{per}}(0, 2\pi)$. We assume that $\lambda = 0$ is a simple eigenvalue of $\mathcal{L}_0$ on $L^2_{\text{per}}(0, 2\pi)$, so that its null space is one-dimensional, and therefore spanned by the derivative $\partial_\theta u_0$ of the wave train.

We may now vary the parameter k in (4.2) near $k = k_0$ and again seek 2π-periodic solutions of (4.2). The derivative of the boundary-value problem (4.2) with respect to ω, evaluated in the solution u_0, is given by $\partial_\theta u_0$. Since $\lambda = 0$ is a simple eigenvalue of $\mathcal{L}_0$ on $L^2_{\text{per}}(0, 2\pi)$, we see that $\partial_\theta u_0$ does not lie in the range of $\mathcal{L}_0$, and the linearization of the boundary-value problem (4.2) with respect to (u, ω) is therefore onto. Thus, exploiting the translation symmetry of (4.2), we can solve uniquely for (u, ω), up to translations in θ, and obtain the wave trains

(4.4) $$u(x,t) = u_0(\omega_{\text{nl}}(k)t - kx; k), \qquad \omega = \omega_{\text{nl}}(k)$$

of (4.1) where $\omega_{\text{nl}}(k_0) = \omega_0$. In particular, $u_0(\theta; k)$ satisfies the ODE

(4.5) $$k^2 D\partial_{\theta\theta} u - \omega_{\text{nl}}(k)\partial_\theta u + f(u) = 0$$

for all k close to k_0. We call ω_{nl} the nonlinear dispersion relation and define the group velocity of the wave train with wave number k to be

(4.6) $$c_{\text{g}} = \frac{\mathrm{d}\omega_{\text{nl}}}{\mathrm{d}k}(k).$$

The phase speed of each wave train is given simply by $c_{\mathrm{p}} := \omega_{\mathrm{nl}}(k)/k$. We shall assume that the nonlinear dispersion relation is genuinely nonlinear so that $\omega_{\mathrm{nl}}''(k_0) \neq 0$.

We will need additional assumptions on the stability of the wave train u_0 as a solution to the reaction-diffusion system (4.1). We therefore linearize (4.1) in the frame $\theta = \omega_0 t - k_0 x$ that moves with the phase speed $c_{\mathrm{p}} = \omega_0/k_0$ of the wave trains and get

$$\partial_t u = k_0^2 D \partial_{\theta\theta} u - \omega_0 \partial_\theta u + f'(u_0(\theta))u. \tag{4.7}$$

The spectrum of the linear operator defined by the right-hand side of (4.7) on the space $L^2(\mathbb{R})$ can be computed using the Floquet or Bloch-wave ansatz

$$u(\theta) = \mathrm{e}^{-\nu\theta/k_0} v(\theta; \nu) \tag{4.8}$$

where $\nu \in \mathrm{i}\mathbb{R}$ and $v(\theta; \nu)$ is 2π-periodic in θ. Substituting this ansatz, we obtain a family of operators $\mathcal{L}_\nu$ given by

$$\mathcal{L}_\nu v = k_0^2 D \left(\partial_\theta - \frac{\nu}{k_0} \right)^2 v - \omega_0 \left(\partial_\theta - \frac{\nu}{k_0} \right) v + f'(u_0(\theta))v, \tag{4.9}$$

each of which is a closed operator on $L^2_{\mathrm{per}}(0, 2\pi)$ with dense domain $H^2_{\mathrm{per}}(0, 2\pi)$. In particular, $\mathcal{L}_\nu$ has compact resolvent, and its spectrum is discrete. Thus, its eigenvalues $\lambda_j(\nu)$ with $j \in \mathbb{N}$ can be ordered with descending real part so that $\operatorname{Re} \lambda_{j+1}(\nu) \leq \operatorname{Re} \lambda_j(\nu)$. The curves $\nu \mapsto \lambda_j(\nu)$ are analytic except possibly at a discrete set where the values of two or more curves $\lambda_j(\nu)$ for different indices j coincide. We denote the associated eigenfunctions by $v_j(\theta; \nu)$.

Since we assumed that $\mathcal{L}_0$ has an algebraically simple eigenvalue at $\lambda = 0$, we find an analytic curve of eigenvalues given by $\lambda = \lambda_{\mathrm{lin}}(\nu)$ for $\nu \in \mathrm{i}\mathbb{R}$ close to zero for which

$$\mathrm{N}(\mathcal{L}_\nu - \lambda_{\mathrm{lin}}(\nu)) \neq \{0\}. \tag{4.10}$$

We call $\lambda = \lambda_{\mathrm{lin}}(\nu)$ the linear dispersion relation. As we shall see below, we can compute the derivative $\mathrm{d}\lambda/\mathrm{d}\nu$ and recover the group velocity as defined via the nonlinear dispersion relation:

$$\left. \frac{\mathrm{d}\lambda_{\mathrm{lin}}}{\mathrm{d}\nu} \right|_{\nu=0} = c_{\mathrm{p}} - \frac{\mathrm{d}\omega_{\mathrm{nl}}}{\mathrm{d}k}(k_0) = c_{\mathrm{p}} - c_{\mathrm{g}}.$$

We remark that the phase velocity c_{p} appears in the above equation simply because we computed the linear dispersion relation in the frame moving with speed c_{p}, while the nonlinear dispersion relation was computed in the steady frame. Note also that the signs of the second derivatives of the linear and nonlinear dispersion relation are, in general, not related. We assume that $\operatorname{Re} \lambda_{\mathrm{lin}}''(0) > 0$.

We summarize the assumptions that we stated so far in the following two hypotheses.

HYPOTHESIS 4.1. *We assume that there exists a wave-train solution to (4.1) whose linearization $\mathcal{L}_0$ has a simple eigenvalue at $\lambda = 0$ when considered on $L^2_{\mathrm{per}}(0, 2\pi)$. We also assume that the nonlinear dispersion relation is genuinely nonlinear so that $\omega_{\mathrm{nl}}''(k_0) \neq 0$*

HYPOTHESIS 4.2. *The linear dispersion relation satisfies $\lambda_{\mathrm{lin}}''(0) > 0$.*

Our last assumption is concerned with temporally oscillatory solutions of the linearization (4.7). We assume that, for each fixed $\nu \in i\mathbb{R}$, and for every λ in the spectrum of $\mathcal{L}_\nu$, we have

(4.11) $$\lambda \neq (c_\mathrm{p} - c_\mathrm{g})\nu$$

except, of course, when $\nu = 0$ and $\lambda = 0$. Recall that c_p and c_g denote the phase and group velocity, respectively, of the wave trains. In the original steady frame, this assumption is equivalent to requiring the absence of solutions to the linearized equation of the form $\exp(i(\omega t - kx))$ for which the phase speed of the modulation, ω/k, is equal to the group velocity c_g of the wave trains. This hypothesis is automatically met if we assume that the wave trains are spectrally stable, see Hypothesis 4.4 below.

HYPOTHESIS 4.3. *We require the absence of resonant spectrum in the dispersion relation as stated in (4.11).*

When we derive and validate the Burgers equation, we shall need the following, more restrictive hypothesis which assumes spectral stability of the wave trains $u_0(\omega_0 t - k_0 x)$.

HYPOTHESIS 4.4. *We assume that, for any $\nu \in i\mathbb{R}$ and any eigenvalue λ of $\mathcal{L}_\nu$, we have either $\mathrm{Re}\,\lambda < 0$ or else $\lambda = 0$ and $\nu \in ik_0\mathbb{Z}$.*

In other words, we assume that $\lambda_\mathrm{lin}(\nu) = \lambda_1(\nu)$, and remark that Hypothesis 4.4 implies Hypothesis 4.3. As indicated above, the stronger Hypothesis 4.4 is only needed for the validity of the Burgers equation and for the stability of small-amplitude shocks but not for their existence.

4.2. Expansions of the linear and nonlinear dispersion relations

In this section, we derive certain useful expansions for the linear and nonlinear dispersion relations. In particular, we show that their first derivatives coincide (up to their sign).

We start with the nonlinear dispersion relation, and consider the nonlinear boundary-value problem (4.2)

(4.12) $$k^2 D \partial_{\theta\theta} u - \omega \partial_\theta u + f(u) = 0$$

with periodic boundary conditions. By assumption, we know that $\lambda = 0$ is a simple eigenvalue of the linearization $\mathcal{L}_0$, posed on $L^2_\mathrm{per}(0, 2\pi)$, of (4.12) about the solution $u_0(\theta)$. The null space of $\mathcal{L}_0$ is therefore spanned by $\partial_\theta u_0$. We denote by u_ad the nontrivial function in the null space of the adjoint operator

$$\mathcal{L}_\mathrm{ad} u = k_0^2 D \partial_{\theta\theta} u + \omega_0 \partial_\theta u + f'(u_0(\theta))^T u$$

with the normalization

(4.13) $$\langle u_\mathrm{ad}, \partial_\theta u_0 \rangle_{L^2(0, 2\pi)} = 1.$$

We now proceed as follows. The above hypothesis on $\mathcal{L}_0$ implies that there is a solution $u_0(\theta; k)$ and $\omega = \omega_\mathrm{nl}(k)$ of (4.12) for each k close to k_0, and that the solution $u \in H^2_\mathrm{per}(0, 2\pi)$ as well as $\omega_\mathrm{nl}(k)$ depend smoothly on k. Thus, we substitute both

$u = u_0(\theta; k)$ and $\omega = \omega_{\mathrm{nl}}(k)$ into (4.12) and take the first two derivatives of (4.12) with respect to k evaluated at $k = k_0$. We obtain

$$\begin{align}
(4.14) \quad \mathcal{L}_0 \partial_k u_0 &= -2k_0 D \partial_{\theta\theta} u_0 + \omega'_{\mathrm{nl}}(k_0) \partial_\theta u_0 \\
(4.15) \quad \mathcal{L}_0 \partial_{kk} u_0 &= -4k_0 D \partial_{\theta\theta} \partial_k u_0 + 2\omega'_{\mathrm{nl}}(k_0) \partial_\theta \partial_k u_0 - 2D \partial_{\theta\theta} u_0 \\
&\quad + \omega''_{\mathrm{nl}}(k_0) \partial_\theta u_0 - f''(u_0)[\partial_k u_0, \partial_k u_0].
\end{align}$$

In particular, we conclude from these equations that the right-hand sides of both (4.14) and (4.15) lie in the range of the operator $\mathcal{L}_0$. Therefore, the L^2-product of these right-hand sides with the adjoint solution u_{ad} vanishes. Writing down these scalar products, and using the normalization (4.13), we obtain

$$\begin{align}
(4.16) \quad \omega'_{\mathrm{nl}}(k_0) &= c_{\mathrm{g}} = \langle u_{\mathrm{ad}}, 2k_0 D \partial_{\theta\theta} u_0 \rangle_{L^2} \\
(4.17) \quad \omega''_{\mathrm{nl}}(k_0) &= \langle u_{\mathrm{ad}}, 4k_0 D \partial_{k\theta\theta} u_0 - 2c_{\mathrm{g}} \partial_k u_0 + 2D \partial_{\theta\theta} u_0 \\
&\quad + f''(u_0)[\partial_k u_0, \partial_k u_0] \rangle_{L^2}.
\end{align}$$

Combining (4.14) and (4.16) shows that $\partial_k u_0$ satisfies

$$(4.18) \quad \mathcal{L}_0 \partial_k u_0 = -2k_0 D \partial_{\theta\theta} u_0 + c_{\mathrm{g}} \partial_\theta u_0 = -2k_0 \left[D \partial_{\theta\theta} u_0 - \langle u_{\mathrm{ad}}, D \partial_{\theta\theta} u_0 \rangle_{L^2} \partial_\theta u_0 \right].$$

We remark that we can arrange for

$$(4.19) \quad \langle u_{\mathrm{ad}}, \partial_k u_0 \rangle_{L^2} = 0$$

since we can always shift wave trains arbitrarily in θ. Shifting appropriately in a k-dependent fashion allows us to replace $\partial_k u_0$ by $\partial_k u_0 + a \partial_\theta u_0$. Choosing a appropriately gives (4.19).

Next, we turn to the linear dispersion relation. We consider the linear boundary-value problem (4.9)

$$(4.20) \quad k_0^2 D \left(\partial_\theta - \frac{\nu}{k_0} \right)^2 v - \omega_0 \left(\partial_\theta - \frac{\nu}{k_0} \right) v + f'(u_0(\theta)) v = \lambda_{\mathrm{lin}}(\nu) v$$

near $\nu = 0$. We proceed as above and take the first two derivatives of this equation with respect to ν evaluated at $\nu = 0$. Upon using that $v = \partial_\theta u_0$ at $\nu = 0$, we obtain

$$\begin{align}
(4.21) \quad \mathcal{L}_0 \partial_\nu v &= (\lambda'_{\mathrm{lin}}(0) - c_{\mathrm{p}}) \partial_\theta u_0 + 2k_0 D \partial_{\theta\theta} u_0 \\
(4.22) \quad \mathcal{L}_0 \partial_{\nu\nu} v &= \lambda''_{\mathrm{lin}}(0) \partial_\theta u_0 + 2\lambda'_{\mathrm{lin}}(0) \partial_\nu v + 4k_0 D \partial_\theta \partial_\nu v \\
&\quad - 2c_{\mathrm{p}} \partial_\nu v - 2D \partial_\theta u_0.
\end{align}$$

Proceeding as above, and comparing the above equations with (4.16) and (4.18), we can arrange that $\partial_\nu v = -\partial_k u_0$, see also (4.19), and

$$\begin{align}
(4.23) \quad \lambda'_{\mathrm{lin}}(0) &= c_{\mathrm{p}} - c_{\mathrm{g}} = \langle u_{\mathrm{ad}}, c_{\mathrm{p}} \partial_\theta u_0 - 2k_0 D \partial_{\theta\theta} u_0 \rangle_{L^2} \\
\lambda''_{\mathrm{lin}}(0) &= \langle u_{\mathrm{ad}}, 4k_0 D \partial_{k\theta} u_0 + 2D \partial_\theta u_0 \rangle_{L^2}.
\end{align}$$

REMARK 4.5. It is sometimes more convenient to use *speed vs period* instead of *spatial vs temporal frequency* formulations of the nonlinear dispersion relation $\omega_{\mathrm{nl}}(k)$. Using

$$(4.24) \quad L = \frac{2\pi}{k}, \quad c_{\mathrm{p}} = \frac{\omega_{\mathrm{nl}}(k)}{k},$$

we obtain the phase velocity $c_{\mathrm{p}} = c(L)$ as a function of L. A trivial application of the chain rule gives

$$(4.25) \quad c_{\mathrm{g}} = \frac{\mathrm{d}\omega_{\mathrm{nl}}}{\mathrm{d}k} = c(L) - L \frac{\mathrm{d}c}{\mathrm{d}L}, \quad \lambda'_{\mathrm{lin}}(0) = L \frac{\mathrm{d}c}{\mathrm{d}L}, \quad \frac{\mathrm{d}^2 \omega_{\mathrm{nl}}}{\mathrm{d}k^2} = \frac{L^3}{2\pi} \frac{\mathrm{d}^2 c}{\mathrm{d}L^2}.$$

Thus, the signs of $-c'(L)$ and the relative group velocity $c_\mathrm{g} - c_\mathrm{p}$ coincide. Furthermore, the signs of $c''(L)$ and $\omega_\mathrm{nl}''(k)$ are the same.

4.3. Formal derivation of the Burgers equation

We are interested in slowly varying modulations of the wave trains $u_0(\omega_\mathrm{nl}(k)t - kx; k)$ to the reaction-diffusion system (4.1)

$$\partial_t u = D\partial_{xx} u + f(u). \tag{4.26}$$

Thus, we fix a wave number k and seek solutions to (4.26) of the form

$$\begin{aligned}u(x,t) &= u_0(\omega_\mathrm{nl}(k)t - kx - \Phi(X,T;\delta); k + \delta\partial_X\Phi(X,T;\delta)) \\ &\quad + \delta^2 u_1(\omega_\mathrm{nl}(k)t - kx, X, T; \delta)\end{aligned} \tag{4.27}$$

where the variables (X, T) depend on (x, t) and are given by

$$X = \delta(x - c_\mathrm{g}(k)t), \qquad T = \delta^2 t. \tag{4.28}$$

We shall assume that both $\Phi(X,T;\delta)$ and $u_1(\theta, X, T; \delta)$ are smooth in δ. The functions $\Phi(X,T;\delta)$ and $q(X,T) := \partial_X \Phi(X,T;\delta)$ describe the slowly varying phase and wave number modulations, respectively.

The strategy is now to derive, on a formal level, the equation that $\Phi(X,T;0)$, or equivalently q, ought to satisfy in order to turn (4.27) into a solution of (4.26). In the process of the derivation, we will also choose a normalization that makes the correction $u_1(\theta, X, T; \delta)$ unique. The validity proof of the equation for $\Phi(X,T;0)$ derived in this fashion then amounts to providing estimates for $\Phi(X,T;\delta)$ and $u_1(\theta, X, T; \delta)$ for $0 < \delta \ll 1$ over time scales of order O(1) in T.

Throughout the derivation, we use the abbreviations

$$\theta = \omega_\mathrm{nl}(k)t - kx, \tag{4.29}$$

$\omega = \omega_\mathrm{nl}(k)$, and $u_0 = u_0(\theta; k)$. For any function $h(\theta - \Phi(X,T;\delta); k + \delta\partial_X\Phi(X,T;\delta))$, we then have

$$\begin{aligned}\frac{\mathrm{d}}{\mathrm{d}t}h &= [\omega + \delta c_\mathrm{g}\partial_X\Phi - \delta^2\partial_T\Phi]\partial_\theta h - \delta^2 c_\mathrm{g}(\partial_{XX}\Phi)\partial_k h + \mathrm{O}(\delta^3) \\ \frac{\mathrm{d}}{\mathrm{d}x}h &= -[k + \delta\partial_X\Phi]\partial_\theta h + \delta^2(\partial_{XX}\Phi)\partial_k h \\ \frac{\mathrm{d}^2}{\mathrm{d}x^2}h &= -\delta^2(\partial_{XX}\Phi)(\partial_\theta h) + [k + \delta\partial_X\Phi]^2\partial_{\theta\theta}h - 2k\delta^2(\partial_{XX}\Phi)\partial_{k\theta}h + \mathrm{O}(\delta^3)\end{aligned}$$

where h is evaluated at $(\theta - \Phi(X,T;\delta); k + \delta\partial_X\Phi(X,T;\delta))$. Therefore, we obtain formally

$$\begin{aligned}&-\partial_t u + D\partial_{xx} u + f(u) \\ &= -(\omega + \delta c_\mathrm{g}\partial_X\Phi - \delta^2\partial_T\Phi)\partial_\theta u_0 + \delta^2 c_\mathrm{g}(\partial_{XX}\Phi)\partial_k u_0 - \delta^2\omega\partial_\theta u_1 \\ &\quad + D\left[-\delta^2(\partial_{XX}\Phi)(\partial_\theta u_0) + (k + \delta\partial_X\Phi)^2\partial_{\theta\theta}u_0 - 2k\delta^2(\partial_{XX}\Phi)\partial_{k\theta}u_0 \right. \\ &\quad \left. + \delta^2 k^2\partial_{\theta\theta}u_1\right] + f(u_0 + \delta^2 u_1) + \mathrm{O}(\delta^3) \\ &= -\omega\partial_\theta u_0 + k^2 D\partial_{\theta\theta}u_0 + f(u_0) + \delta\left[2k(\partial_X\Phi)D\partial_{\theta\theta}u_0 - c_\mathrm{g}(\partial_X\Phi)\partial_\theta u_0\right] \\ &\quad + \delta^2\left[c_\mathrm{g}(\partial_{XX}\Phi)\partial_k u_0 + (\partial_T\Phi)\partial_\theta u_0 \right. \\ &\quad + D\left(-(\partial_{XX}\Phi)\partial_\theta u_0 + (\partial_X\Phi)^2\partial_{\theta\theta}u_0 - 2k(\partial_{XX}\Phi)\partial_{k\theta}u_0\right) \\ &\quad \left. -\omega\partial_\theta u_1 + k^2 D\partial_{\theta\theta}u_1 + f'(u_0)u_1\right] + \mathrm{O}(\delta^3)\end{aligned}$$

where $u_0 = u_0(\theta - \Phi(X,T;0); k + \delta\partial_X\Phi(X,T;0))$ and $u_1 = u_1(\theta, X, T; 0)$. In the next step, we expand u_0 further

$$\begin{aligned}
u_0(\theta - \Phi; k + \delta\partial_X\Phi) &= u_0(\theta - \Phi; k) + \delta(\partial_X\Phi)\partial_k u_0(\theta - \Phi; k) \\
&\quad + \frac{\delta^2}{2}(\partial_X\Phi)^2 \partial_k^2 u_0(\theta - \Phi; k) + \mathrm{O}(\delta^3)
\end{aligned}$$

and note that analogous expansions hold for $\partial_\theta u_0$ and $\partial_{\theta\theta} u_0$. We will next substituting these expansions into the above equation for $-\partial_t u + D\partial_{xx} u + f(u)$. From this point on, we regard (θ, X, T) as independent variables and neglect that they depend on x and t through (4.28) and (4.29). We shall also set

$$\Phi(X, T) := \Phi(X, T; 0).$$

Upon substitution, we obtain, with $u_0 = u_0(\theta - \Phi(X,T); k)$ and $u_1 = u_1(\theta, X, T; 0)$, that

$$\begin{aligned}
&-\partial_t u + D\partial_{xx} u + f(u) \\
&= -\omega\partial_\theta u_0 + k^2 D\partial_{\theta\theta} u_0 + f(u_0) \\
&\quad + \delta(\partial_X\Phi)[-\omega\partial_{k\theta} u_0 + k^2 D\partial_{k\theta\theta} u_0 + f'(u_0)\partial_k u_0 + 2kD\partial_{\theta\theta} u_0 - c_g\partial_\theta u_0] \\
&\quad + \frac{\delta^2}{2}[2c_g(\partial_{XX}\Phi)\partial_k u_0 - \omega(\partial_X\Phi)^2 \partial_{kk\theta} u_0 + k^2 D(\partial_X\Phi)^2 \partial_{kk\theta\theta} u_0 \\
&\qquad + (\partial_X\Phi)^2 f'(u_0)\partial_{kk} u_0 + (\partial_X\Phi)^2 f''(u_0)[\partial_k u_0, \partial_k u_0] + 4k(\partial_X\Phi)^2 D\partial_{k\theta\theta} u_0 \\
&\qquad - 2c_g(\partial_X\Phi)^2 \partial_{k\theta} u_0 + 2(\partial_T\Phi)\partial_\theta u_0 + 2D(-(\partial_{XX}\Phi)\partial_\theta u_0 + (\partial_X\Phi)^2 \partial_{\theta\theta} u_0 \\
&\qquad + 2k(\partial_{XX}\Phi)\partial_{k\theta} u_0) + \mathcal{L}_0 u_1] + \mathrm{O}(\delta^3) \\
&= -\omega\partial_\theta u_0 + k^2 D\partial_{\theta\theta} u_0 + f(u_0) \\
&\quad + \delta(\partial_X\Phi)[\mathcal{L}_0 \partial_k u_0 + 2kD\partial_{\theta\theta} u_0 - c_g\partial_\theta u_0] \\
&\quad + \frac{\delta^2}{2}[2\mathcal{L}_0 u_1 + (\partial_X\Phi)^2(\mathcal{L}_0 \partial_{kk} u_0 + f''(u_0)[\partial_k u_0, \partial_k u_0] + 4kD\partial_{k\theta\theta} u_0 \\
&\qquad - 2c_g\partial_{k\theta} u_0 + 2D\partial_{\theta\theta} u_0) + 2(\partial_T\Phi)\partial_\theta u_0 - 2D(\partial_{XX}\Phi)(\partial_\theta u_0 + 2k\partial_{k\theta} u_0) \\
&\qquad + 2c_g(\partial_{XX}\Phi)\partial_k u_0] + \mathrm{O}(\delta^3)
\end{aligned}$$

where $\mathcal{L}_0$ has been defined in (4.3). Equations (4.12) and (4.14) imply that the terms of order $\mathrm{O}(\delta^0)$ and $\mathrm{O}(\delta)$ in the above expression vanish identically. We can then use (4.15) to simplify the term of order $\mathrm{O}(\delta^2)$ to get

$$\begin{aligned}
&-\partial_t u + D\partial_{xx} u + f(u) \\
&= \delta^2\Big[\mathcal{L}_0 u_1 + \frac{1}{2}\omega''_{\mathrm{nl}}(k)(\partial_X\Phi)^2 \partial_\theta u_0 + (\partial_T\Phi)\partial_\theta u_0 - D(\partial_{XX}\Phi)(\partial_\theta u_0 + 2k\partial_{k\theta} u_0) \\
&\quad + c_g(\partial_{XX}\Phi)\partial_k u_0\Big] + \mathrm{O}(\delta^3).
\end{aligned}$$

Thus, we require that the term of order $\mathrm{O}(\delta^2)$ vanishes identically which results in

$$(4.30) \quad \begin{aligned}
\mathcal{L}_0 u_1 &= -(\partial_T\Phi)\partial_\theta u_0 - \frac{1}{2}\omega''_{\mathrm{nl}}(k)(\partial_X\Phi)^2 \partial_\theta u_0 \\
&\quad + D(\partial_{XX}\Phi)(\partial_\theta u_0 + 2k\partial_{k\theta} u_0) + c_g(\partial_{XX}\Phi)\partial_k u_0.
\end{aligned}$$

We can solve this equation uniquely for $u_1 = u_1(\theta, X, T; 0)$ provided we require that $\langle u_{\mathrm{ad}}, u_1\rangle_{L^2} = 0$ and provided we choose $\Phi(X, T)$ so that the right-hand side is in the range of $\mathcal{L}_0$, that is, provided

$$\left\langle u_{\mathrm{ad}}, -(\partial_T \Phi)\partial_\theta u_0 - \frac{1}{2}\omega''_{\mathrm{nl}}(k)(\partial_X \Phi)^2 \partial_\theta u_0 + D(\partial_{XX}\Phi)(\partial_\theta u_0 + 2k\partial_{k\theta}u_0) \right.$$
$$\left. + c_{\mathrm{g}}(\partial_{XX}\Phi)\partial_k u_0 \right\rangle_{L^2} = 0$$

Exploiting (4.19), this solvability condition becomes

$$\langle u_{\mathrm{ad}}, \partial_\theta u_0 \rangle_{L^2} \partial_T \Phi$$
$$= \langle u_{\mathrm{ad}}, D(\partial_\theta u_0 + 2k\partial_{k\theta}u_0)\rangle_{L^2} \partial_{XX}\Phi - \frac{\omega''_{\mathrm{nl}}(k)}{2} \langle u_{\mathrm{ad}}, \partial_\theta u_0 \rangle_{L^2} (\partial_X \Phi)^2.$$

Using (4.13) and (4.23), we eventually obtain the eikonal equation

$$(4.31) \qquad \partial_T \Phi = \frac{\lambda''_{\mathrm{lin}}(0)}{2}\partial_{XX}\Phi - \frac{\omega''_{\mathrm{nl}}(k)}{2}(\partial_X \Phi)^2$$

for the phase $\Phi(X,T)$ or, alternatively, the Burgers equation

$$(4.32) \qquad \partial_T q = \frac{\lambda''_{\mathrm{lin}}(0)}{2}\partial_{XX}q - \frac{\omega''_{\mathrm{nl}}(k)}{2}\partial_X(q^2)$$
$$= \frac{\lambda''_{\mathrm{lin}}(0)}{2}\partial_{XX}q - \omega''_{\mathrm{nl}}(k)\, q\partial_X q$$

for the wave number modulation $q(X,T) = \partial_X \Phi(X,T)$.

4.4. Validity of the Burgers equation

We start with the wave trains

$$u(x,t) = u_0(\omega_{\mathrm{nl}}(k)t - kx; k)$$

of the reaction-diffusion system (4.1) for a fixed wave number k. It turns out to be convenient to replace the spatial variable x by $\theta = \omega_{\mathrm{nl}}(k)t - kx$. In these coordinates, (4.1) becomes

$$(4.33) \qquad \partial_t u = k^2 D \partial_{\theta\theta} u - \omega_{\mathrm{nl}}(k)\partial_\theta u + f(u),$$

and the wave trains are simply given by

$$(4.34) \qquad u(\theta,t) = u_0(\theta; k).$$

To get into the spirit of the results we shall prove, we will begin with statements that are relatively easy to formulate but may not be the most general or relevant ones (the latter ones can be found toward the end of this section).

Therefore, we shall first consider slowly-varying modulations of the wave trains (4.34) that admit the following representation. Pick a phase function $\phi(\vartheta,t)$ with $|\partial_\vartheta \phi(\vartheta,t)| \leq 1/2$ uniformly in (ϑ,t) and consider the change of coordinates defined implicitly via

$$(4.35) \qquad \theta = \vartheta + \phi(\vartheta,t).$$

Due to our assumption on ϕ, we can solve (4.35) for ϑ as a function $\vartheta(\theta)$ of θ, which will allow us to write solutions of (4.33) in the form $u(\theta,t) = U(\vartheta,t)$, where ϑ and θ are related via (4.35).

Initially, for the sake of clarity, we shall formulate results in terms of the variables ϑ; results for the original variable θ are stated toward the end of this section.

First, we consider modulation of the phase. Thus, pick a solution $\Phi(X,T)$ of the phase equation

(4.36) $$\partial_T \Phi = \frac{1}{2}\lambda''_{\text{lin}}(0)\partial_{XX}\Phi - \frac{1}{2}\omega''_{\text{nl}}(k)(\partial_X \Phi)^2$$

with $X \in \mathbb{R}$ and $T \in [0, T_0]$, and set

(4.37) $$\phi(\vartheta, t) := \Phi\left(\delta((c_{\text{p}} - c_{\text{g}})t - \vartheta/k), \delta^2 t\right).$$

We then have the following approximation result.

THEOREM 4.6. *Assume that Hypotheses 4.1, 4.2 and 4.4 are met, and fix an integer $n \geq 3$. For each choice of constants $C_0 > 0$ and $T_0 > 0$, there exist constants $\delta_1 > 0$ and $C_1 > 0$ such that the following is true: Pick a solution $\Phi(X,T)$ of (4.36) with*

$$\sup_{T \in [0,T_0]} \|\Phi(\cdot, T)\|_{H^n} \leq C_0$$

and define

$$U_{\text{approx}}(\vartheta, t) = u_0\left(\vartheta; k(1 + \delta \partial_X \Phi(X, T))\right), \qquad (X, T) := \left(\delta((c_{\text{p}} - c_{\text{g}})t - \vartheta/k), \delta^2 t\right)$$

then there exists a solution $u(\theta, t) = U(\vartheta, t)$ of the reaction-diffusion system (4.33) such that

$$\sup_{t \in [0, T_0/\delta^2]} \sup_{\vartheta \in \mathbb{R}} |U(\vartheta, t) - U_{\text{approx}}(\vartheta, t)| \leq C_1 \delta^2$$

uniformly in $\delta \in (0, \delta_1)$, where the variables ϑ and θ are related through (4.35) and (4.37).

The preceding theorem is weaker than Theorem 3.2 which we stated for the complex Ginzburg–Landau equation in §3.2, since Φ has to lie in H^n instead of H^n_{ul}. The reason is of a technical nature: we do not know of estimates for quadratic interactions of H^n_{ul}-functions which retain the scaling with respect to the Bloch variable ℓ, while such estimates exist for H^n-functions [54].

Solutions with a richer dynamics can be obtained by modulating the wave number instead of the phase, and we are therefore interested in solutions $q(X,T) = \partial_X \Phi_0(X,T)$ of the Burgers equation

(4.38) $$\partial_T q = \frac{1}{2}\lambda''_{\text{lin}}(0)\partial_{XX} q - \frac{1}{2}\omega''_{\text{nl}}(k)\partial_X(q^2).$$

In this situation, we have the following result.

THEOREM 4.7. *We assume that Hypotheses 4.1, 4.2 and 4.4 are met. For each choice of integers $M \geq 1$ and $n \geq M + 3$ and constants $C_0 > 0$ and $T_0 > 0$, there exist constants $\delta_1 > 0$ and $C_1 > 0$ with the following properties: For each solution $q(X,T)$ of the Burgers equation (4.38) for which*

$$\sup_{T \in [0,T_0]} \|q(\cdot, T)\|_{H^n_{\text{ul}}} \leq C_0$$

and each $\delta \in (0, \delta_1)$, there are functions $(q_h, r_h)(\vartheta, t)$ with

$$\sup_{t \in [0, T_0/\delta^2]} \left\|q_h(\cdot, t) - q\left(\delta((c_{\text{p}} - c_{\text{g}})t - \vartheta/k), \delta^2 t\right)\right\|_{H^n_{\text{ul}}} \leq C_1 \delta$$

$$\sup_{t \in [0, T_0/\delta^2]} \|r_h(\cdot, t)\|_{H^n_{\text{ul}}} \leq C_1$$

and a solution $u(\theta, t) = U(\vartheta, t)$ of the reaction-diffusion system (4.33) such that
$$\sup_{t \in [0, T_0/\delta^2]} \sup_{\vartheta \in \mathbb{R}} |U(\vartheta, t) - U_{\text{approx}}(\vartheta, t)| \leq C_1 \delta^{1+M},$$
where
$$U_{\text{approx}}(\vartheta, t) = u_0\left(\vartheta; k(1 + \delta q_h(\vartheta, t))\right) + \delta^2 r_h(\vartheta, t).$$
Again, ϑ and θ are related through (4.35) with
$$\phi(\vartheta, t) := \delta \int_0^\vartheta q_h(\tilde{\vartheta}, t) \, d\tilde{\vartheta}.$$

The functions $(q_h, r_h)(x, t; \delta)$ are obtained as higher-order approximations to the solution $q(X, T)$ of the Burgers equation. As outlined in §4.3, they can, in principle, be computed by solving a recursive set of linear inhomogeneous PDEs.

For practical purposes, the approximation results stated so far are quite useless as they are formulated in terms of the coordinate ϑ which is defined implicitly by $\theta = \vartheta + \phi(\vartheta, t)$, an expression that obviously involves the knowledge of the phase function ϕ. Hence, we now transfer our assertions from the ϑ to the θ variable.

First, we state an approximation result for solutions $q(X, T)$ to (4.38) that approach different constants $q_\pm$ as $X \to \pm\infty$. Such solutions are of particular interest, since they describe the temporal evolution of interfaces between wave trains with wave numbers $k + \delta q_\pm$ at $X = \pm\infty$.

THEOREM 4.8. *Assume that Hypotheses 4.1, 4.2 and 4.4 are met, and fix integers $M \geq 3$ and $n \geq M + 3$. For each choice of $C_0 > 0$ and $T_0 > 0$, there exist constants $\delta_1 > 0$ and $C_1 > 0$ such that the following is true: Pick a $\delta \in (0, \delta_1)$ and a solution $q(X, T)$ of the Burgers equation (4.38) for which there are numbers $q_\pm \in \mathbb{R}$ with*
$$\sup_{T \in [0, T_0]} \left[\|q(\cdot, T)\|_{H_{\text{ul}}^n} + \|(q(\cdot, T) - q_+)\rho_{\text{w}}^2\|_{H_{\text{ul}}^n(\mathbb{R}^+)} + \|(q(\cdot, T) - q_-)\rho_{\text{w}}^2\|_{H_{\text{ul}}^n(\mathbb{R}^-)} \right] \leq C_0$$
where $\rho_{\text{w}}(X) = \sqrt{1 + X^2}$. Then there are functions $(q_h, r_h)(\vartheta, t)$ with
$$\sup_{t \in [0, T_0/\delta^2]} \left\| q_h(\cdot, t) - q\left(\delta((c_{\text{p}} - c_{\text{g}})t - \vartheta/k), \delta^2 t\right) \right\|_{H_{\text{ul}}^n} \leq C_1 \delta$$
$$\sup_{t \in [0, T_0/\delta^2]} \|r_h(\cdot, t)\|_{H_{\text{ul}}^n} \leq C_1$$
and a solution $u(\theta, t)$ of the reaction-diffusion system (4.33) such that
$$\sup_{t \in [0, T_0/\delta^2]} \sup_{\theta \in \mathbb{R}} |u(\theta, t) - u_{\text{approx}}(\theta, t)| \leq C_1 \delta^{M-3/2},$$
where u_{approx} is given by
$$u_{\text{approx}}(\theta, t) = u_0\left(\vartheta(\theta); k(1 + \delta q_h(\vartheta(\theta), t))\right) + \delta^2 r_h(\vartheta(\theta), t),$$
and $\vartheta(\theta)$ is the solution of
$$\theta = \vartheta + \phi(\vartheta, t), \qquad \phi(\vartheta, t) := \delta q_- \vartheta + \delta \int_{-\infty}^\vartheta \left(q_h(\tilde{\vartheta}, t) - q_-\right) d\tilde{\vartheta}.$$

Lastly, we state the most general approximation result that we were able to prove. We will encounter the same limitations that we found earlier for the complex Ginzburg–Landau equation: We cannot expect validity to hold uniformly in $\theta \in \mathbb{R}$ but only for θ in bounded intervals where the length of the interval depends on the accuracy of the initial approximation. Furthermore, we need to add a global x-independent phase shift $\phi_0(t)$ which will be of order $O(1)$ in δ: therefore, only the profile of modulations but not their exact location is approximated over the relevant natural time scales.

THEOREM 4.9. *Assume that Hypotheses 4.1, 4.2 and 4.4 are met, and fix integers n, M and a real number l with $M \geq 1$, $n \geq M + 3$ and $0 < l < M$. For each choice of $C_0 > 0$ and $T_0 > 0$, there exist constants $\delta_1 > 0$ and $C_1 > 0$ such that the following is true: Pick a solution $q(X, T)$ of the Burgers equation (4.38) for which*

$$\sup_{T \in [0, T_0]} \|q(\cdot, T)\|_{H^n_{\mathrm{ul}}} \leq C_0$$

and $\delta \in (0, \delta_1)$, then there are functions $(q_h, r_h) \in H^n_{\mathrm{ul}}$ with

$$\sup_{t \in [0, T_0/\delta^2]} \left\| q_h(\cdot, t) - q\left(\delta((c_{\mathrm{p}} - c_{\mathrm{g}})t - \vartheta/k), \delta^2 t\right) \right\|_{H^n_{\mathrm{ul}}} \leq C_1 \delta$$

$$\sup_{t \in [0, T_0/\delta^2]} \|r_h(\cdot, t)\|_{H^n_{\mathrm{ul}}} \leq C_1,$$

a phase function $\phi_0(t)$ with

$$\sup_{t \in [0, T_0/\delta^2]} |\phi_0(t)| \leq C_1,$$

and a solution $u(\theta, t)$ of the reaction-diffusion system (4.33) such that

$$\sup_{t \in [0, T_0/\delta^2]} \sup_{|\theta| \leq L/\delta^l} |u(\theta, t) - u_{\mathrm{approx}}(\theta, t)| \leq C_1 \delta^{1+M-l},$$

where u_{approx} is given by

$$u_{\mathrm{approx}}(\theta, t) = u_0\left(\vartheta(\theta); k(1 + \delta q_h(\vartheta(\theta), t))\right) + \delta^2 r_h(\vartheta(\theta), t),$$

and $\vartheta(\theta)$ is the solution of

$$\theta = \vartheta + \phi(\vartheta, t), \qquad \phi(\vartheta, t) := \phi_0(t) + \delta \int_0^\vartheta q_h(\tilde\vartheta, t) \, \mathrm{d}\tilde\vartheta.$$

The proofs of the theorems stated in this section can be found in §5.

4.5. Existence and stability of weak shocks

We are interested in solutions to (4.1) that are spatially bi-asymptotic to wave trains and time-periodic with temporal frequency ω_* in a frame moving with an appropriate speed c_*. In other words, we seek solutions $u(x,t) = u_*(x - c_* t, \omega_* t)$ such that

(4.39) $\quad u_*(x - c_* t, \omega_* t) \;=\; u_*(x - c_* t, \omega_* t + 2\pi)$

(4.40) $\quad u_*(x - c_* t, \omega_* t) \;\longrightarrow\; u_0(\omega_\pm t - k_\pm x; k_\pm) \qquad$ as $x \to \pm\infty$.

Here, the convergence is understood to be uniformly in t for u and its derivatives $\partial_x u$ and $\partial_t u$. More precisely, upon using the new independent variables $\xi = x - c_* t$ and $\tau = \omega_* t$, we require that

(4.41) $\left\| u_*(\xi, \cdot) - u_0\left(\dfrac{\omega_\pm - k_\pm c_*}{\omega_*} \cdot - k_\pm \xi; k_\pm\right) \right\|_{H^1_{\mathrm{per}}(0, 2\pi)}$
$+ \left\| \partial_\xi u_*(\xi, \cdot) - \partial_\xi u_0\left(\dfrac{\omega_\pm - k_\pm c_*}{\omega_*} \cdot - k_\pm \xi; k_\pm\right) \right\|_{H^{1/2}_{\mathrm{per}}(0, 2\pi)} \longrightarrow 0$

4.5. EXISTENCE AND STABILITY OF WEAK SHOCKS

as $\xi \to \pm\infty$. The speed c_* and the asymptotic wave numbers $k_\pm$ are, in principle, free parameters. Note, however, that the asymptotic frequencies $\omega_\pm$ are necessarily fixed through the nonlinear dispersion relation $\omega_\pm = \omega_{\mathrm{nl}}(k_\pm)$. In particular, since (4.39) requires that the solution has frequency one in τ, we see that the asymptotics required in (4.41) imply that

$$\frac{\omega_{\mathrm{nl}}(k_+) - k_+ c_*}{\omega_*} = 1 = \frac{\omega_{\mathrm{nl}}(k_-) - k_- c_*}{\omega_*}$$

so that

(4.42) $$\omega_{\mathrm{nl}}(k_+) - k_+ c_* = \omega_* = \omega_{\mathrm{nl}}(k_-) - k_- c_*.$$

Hence, for $k_- \neq k_+$, we find that the average speed c_* of our solution is determined by the Rankine–Hugoniot condition

(4.43) $$c_* = \frac{\omega_{\mathrm{nl}}(k_+) - \omega_{\mathrm{nl}}(k_-)}{k_+ - k_-}$$

with the genuinely nonlinear flux function ω_{nl}. Using (4.42), we see that the corresponding frequency ω_* is then given by

(4.44) $$\omega_* = \frac{k_+ \omega_{\mathrm{nl}}(k_-) - k_- \omega_{\mathrm{nl}}(k_+)}{k_+ - k_-}.$$

We have $c_* \to c_{\mathrm{g}}$ and $\omega_* \to k_0(c_{\mathrm{p}} - c_{\mathrm{g}})$ as $k_+, k_- \to k_0$. We say that the solution is a viscous shock if

(4.45) $$c_{\mathrm{g}}^- > c_* > c_{\mathrm{g}}^+$$

where $c_{\mathrm{g}}^\pm = \omega_{\mathrm{nl}}'(k_\pm)$ or, equivalently, if

(4.46) $$\omega_{\mathrm{nl}}'(k_-) > \frac{\omega_{\mathrm{nl}}(k_+) - \omega_{\mathrm{nl}}(k_-)}{k_+ - k_-} > \omega_{\mathrm{nl}}'(k_+).$$

In particular, viscous shocks correspond to the viscous Lax shocks of the Burgers equation found in §2.2. The interpretation of (4.45) is that the asymptotic wave trains transport toward the interface between them. For $k_\pm$ close to k_0, (4.46) implies that $k_- < k_0 < k_+$ for concave dispersion relations with $\omega_{\mathrm{nl}}''(k_0) < 0$, while we have $k_+ < k_0 < k_-$ for convex dispersion relations with $\omega_{\mathrm{nl}}''(k_0) > 0$ (see also Figure 1.3).

THEOREM 4.10 (Existence). *Assume that Hypotheses 4.1, 4.2 and 4.3 are met, then the following is true for all wave numbers k_- and k_+ that are sufficiently close to k_0 and for which $c_{\mathrm{g}}^- > c_{\mathrm{g}}^+$. There exists a viscous shock solution $u(x,t) = u_*(x - c_* t, \omega_* t)$ of (4.1) whose temporal frequency ω_* and speed c_* are determined by the Rankine–Hugoniot conditions (4.43)-(4.44). Furthermore, the viscous shock is unique, up to translations in x and t, in the class of solutions that are close to u_0 and satisfy (4.39)-(4.40).*

REMARK 4.11. We emphasize that the existence statement of Theorem 4.10 remains true if $\lambda_{\mathrm{lin}}''(0) < 0$ in Hypothesis 4.2. The resulting modulated waves, however, are sources, and not shocks, and satisfy $c_{\mathrm{g}}^- < c_* < c_{\mathrm{g}}^+$. The rest of Theorem 4.10 remains true as stated.

Under slightly more restrictive assumptions on the linear dispersion relation, the small-amplitude viscous shocks $u_*(\xi, \tau)$ that we found in the preceding theorem are spectrally stable.

We formulate the stability result in exponentially weighted spaces. For each η_- and $\eta_+ \in \mathbb{R}$, we define

$$
(4.47) \quad L^2_{\eta_-,\eta_+}(\mathbb{R}) := \left\{ u \in L^2_{\mathrm{loc}}(\mathbb{R});\ \|u\|_{L^2_{\eta_-,\eta_+}} < \infty \right\}
$$

$$
\|u\|^2_{L^2_{\eta_-,\eta_+}} := \int_{-\infty}^0 |u(x)\mathrm{e}^{\eta_- x}|^2\, \mathrm{d}x + \int_0^\infty |u(x)\mathrm{e}^{\eta_+ x}|^2\, \mathrm{d}x.
$$

We also define $H^1_{\eta_-,\eta_+}$ as the subset of functions u in H^1_{loc} for which both u and u_x belong to $L^2_{\eta_-,\eta_+}$.

THEOREM 4.12 (Stability). *Assume that Hypotheses 4.1, 4.2 and 4.4 are met. Let $u_*(x - c_* t, \omega_* t)$ be the viscous shock solution found in Theorem 4.10 with asymptotic wave numbers $k_\pm$. There exist then extremal weights $\eta_{\min}(k_\pm)$ and $\eta_{\max}(k_\pm)$ such that, for each fixed choice of $\eta_\pm$ with $\eta_{\min} < \eta_- < 0 < \eta_+ < \eta_{\max}$, the viscous shock is nonlinearly asymptotically stable with respect to perturbations in $H^1_{\eta_-,\eta_+}$. More precisely, fix $\eta_\pm$ with $\eta_{\min} < \eta_- < 0 < \eta_+ < \eta_{\max}$, then there are positive numbers δ, ρ and C such that each solution $u(x,t)$ of (4.1) for which $\|u(\cdot,0) - u_*(\cdot,0)\|_{H^1_{\eta_-,\eta_+}} < \delta$ satisfies*

$$
(4.48) \quad \|u(\cdot,t) - u_*(\cdot - c_* t, \omega_* t)\|_{H^1_{\eta_-,\eta_+}} \leq C \mathrm{e}^{-\rho t} \|u(\cdot,0) - u_*(\cdot,0)\|_{H^1_{\eta_-,\eta_+}}.
$$

The optimal exponential rate of convergence is given by

$$
\rho = \min\{|(c_\mathrm{g}^- - c_*)\eta_-|, |(c_\mathrm{g}^+ - c_*)\eta_+|\}.
$$

The theorem asserts that localized perturbations do not cause a shift in the position or the phase of the front. In particular, the linearization of the reaction-diffusion system about the viscous shock solution does not have a neutral eigenvalue at the origin when considered in the exponentially weighted spaces used in the theorem. Note also that stronger exponential weights enhance the exponential rate with which perturbations decay in time.

At a first glance, this statement may appear to contradict the assertion in Proposition 2.4 that the viscous Lax shocks in the Burgers equation do have a neutral eigenvalue at the origin in the same exponentially weighted spaces. The explanation is as follows. The exponential weights require that perturbations of the viscous shocks in the reaction-diffusion system are localized. In particular, the phases of the asymptotic wave trains cannot be changed by perturbations in these spaces. The phase modulation is modelled by the eikonal equation

$$
(4.49) \quad \partial_T \Phi = \frac{1}{2} \lambda''_{\mathrm{lin}}(0) \partial_{XX} \Phi - \frac{1}{2} \omega''_{\mathrm{nl}}(k)(\partial_X \Phi)^2
$$

for the phase $\Phi = \int q$, i.e. by the integrated Burgers equation, instead of the Burgers equation for the wave number q. The linearization of (4.49) about the Lax shock does not possess a zero eigenvalue in our exponentially weighted spaces since $\Phi_x = q$ is not localized. Equivalently, we can view perturbations of (4.49) that are localized in Φ as perturbations of the Burgers equation that have zero mass $\int q = 0$: these perturbations, however, preserve the position of the viscous shock. The absence of the zero eigenvalue can also be inferred directly from the properties of the viscous shock in the reaction-diffusion system. Any eigenfunction of $\lambda = 0$ for the linearization of the reaction-diffusion system about the shock solution is a linear combination of the derivatives of the shock solution with respect to time and space; we would obtain an eigenvalue at $\lambda = 0$ in the exponentially weighted spaces considered above if an appropriate linear combination is localized in space. We shall prove, however, that localized linear combinations do not exist: Moving the interface in space and time will always change the phases of at least one of the asymptotic wave trains since phase speeds of the two wave trains differ (see also [50, Proposition 5.5] where it is shown that the absence of zero eigenvalues is a generic feature of sinks regardless of their amplitude).

The proofs of the theorems stated in this section can be found in §8.

CHAPTER 5

Validity of the Burgers equation in reaction-diffusion equations

In this section, we prove the theorems that we stated in §4.4. In §5.1–5.4, we carry out the proof of Theorem 4.7, while we discuss in §5.5 the modifications needed for the other theorems.

5.1. From phases to wave numbers

The first step is to extract an equation for the phase, and henceforth for the wave number, from a general reaction-diffusion system. Note that the formal derivation in §4.3 uses a multi-scale expansion which we cannot assume apriori. We remark that the S^1-symmetry respected by the Ginzburg–Landau equation provided an avenue for deriving directly an equation for the phase. This symmetry, however, is, in general, not respected in reaction-diffusion systems.

Instead, we proceed as follows for the reaction-diffusion system

(5.1) $$\partial_t u = D\partial_{xx} u + f(u).$$

First, we change coordinates via

$$\theta = \omega t - kx$$

and obtain

(5.2) $$\partial_t u = k^2 D\partial_{\theta\theta} u - \omega \partial_\theta u + f(u).$$

Our starting point is a given stationary wave train $u_0(\theta; k)$ of (5.2) with period 2π, which therefore satisfies

(5.3) $$k^2 D\partial_{\theta\theta} u_0 - \omega \partial_\theta u_0 + f(u_0) = 0.$$

Given a smooth phase function $\phi(\vartheta, t)$, we shall now seek solutions of the form

(5.4) $$u(\theta, t) = u_0(\vartheta; k(1 + \partial_\vartheta \phi(\vartheta, t))^{-1}) + w(\vartheta, t),$$

where the phase $\phi(\vartheta, t)$ and the coordinates θ and ϑ are related by

(5.5) $$\theta = \vartheta + \phi(\vartheta, t).$$

We shall assume that $\partial_\vartheta \phi$ is small, uniformly in ϑ, but remark that ϕ itself might be unbounded.

Through the ansatz (5.4), we add an additional degree of freedom by introducing ϕ: we shall later add additional conditions on ϕ and w, via mode filters, to make the decomposition (5.4) unique.

REMARK 5.1. It might seem more natural to make the ansatz

(5.6) $$u(\theta, t) = u_0(\theta - \phi(\theta, t); k(1 - \partial_\theta \phi(\theta, t))) + w(\theta, t)$$

54 5. VALIDITY OF THE BURGERS EQUATION IN REACTION-DIFFUSION EQUATIONS

instead of (5.4). Eventually, we need to be able to relate the dynamics of $u(\theta, t)$ back to properties of the wave train $u_0(\theta; k)$. It is desirable to allow the phase function $\phi(\theta, t)$ to be unbounded. Thus, to transform back to the wave train, we would need to express $u(\theta, t)$ in terms of the variable $\vartheta = \theta - \phi(\theta, t)$ which yields

$$u_0(\theta - \phi(\theta, t); k(1 - \partial_\theta \phi(\theta, t))) \longmapsto u_0(\vartheta; k(1 - \partial_\theta \phi(\theta(\vartheta, t), t)))$$

which involves the inverse $\theta(\vartheta, t)$ of the function $\vartheta = \theta - \phi(\theta, t)$. The occurrence of this inverse would make the forthcoming analysis much more complicated which is why we proceed with (5.4).

REMARK 5.2. Suppose that we found a phase function $\phi(\vartheta, t)$ with small derivative $\partial_\vartheta \phi(\vartheta, t)$ so that (5.4) satisfies (5.2). Using the implicit function theorem, we can then, a posteriori, solve (5.5) for ϑ as a function of θ which is of the form $\vartheta = \theta - \tilde{\phi}(\theta, t)$, where

$$\tilde{\phi}(\theta, t) = \phi(\vartheta, t) = \phi(\theta - \tilde{\phi}(\theta, t), t).$$

In particular, we see that $u_0(\vartheta; k(1 + \partial_\theta \phi(\vartheta, t))^{-1})$ becomes

(5.7) $$u_0(\theta - \phi(\theta - \tilde{\phi}(\theta, t), t); k(1 + \partial_\theta \phi(\theta - \tilde{\phi}(\theta, t), t))^{-1})$$

and

$$\begin{aligned}\frac{\mathrm{d}}{\mathrm{d}\theta}\phi(\theta - \tilde{\phi}(\theta, t), t) &= (1 - \partial_\theta \tilde{\phi}(\theta, t))\partial_\theta \phi(\theta - \tilde{\phi}(\theta, t), t) \\ &= \partial_\theta \phi(\theta - \tilde{\phi}(\theta, t), t) + \mathrm{O}(|\partial_\theta \phi(\theta - \tilde{\phi}(\theta, t), t)|^2).\end{aligned}$$

Thus, to leading order, the solution (5.7) is, in fact, of the desired form (5.6) with $\phi(\theta, t)$ replaced by $\phi(\theta - \tilde{\phi}(\theta, t), t)$.

We now substitute the ansatz (5.4) into (5.2) and derive the resulting PDE in ϑ. We shall use the notation

(5.8) $$\begin{aligned}u_0^\phi &:= u_0(\vartheta; k(1 + \partial_\vartheta \phi)^{-1}) \\ \partial_j u_0^\phi &:= (\partial_j u_0)^\phi := (\partial_j u_0)(\vartheta; k(1 + \partial_\vartheta \phi)^{-1}), \qquad j = \vartheta, k.\end{aligned}$$

Assuming that $\partial_\vartheta \phi$ is small, we obtain the relations

$$\frac{\mathrm{d}\vartheta}{\mathrm{d}t} = \frac{-\partial_t \phi}{1 + \partial_\vartheta \phi}$$

$$\frac{\mathrm{d}\vartheta}{\mathrm{d}\theta} = \frac{1}{1 + \partial_\vartheta \phi}$$

$$\frac{\mathrm{d}}{\mathrm{d}\theta} = \frac{1}{1 + \partial_\vartheta \phi}\frac{\mathrm{d}}{\mathrm{d}\vartheta}$$

$$\frac{\mathrm{d}^2}{\mathrm{d}\theta^2} = \left(\frac{1}{1 + \partial_\vartheta \phi}\frac{\mathrm{d}}{\mathrm{d}\vartheta}\right)^2$$

and therefore

$$\frac{\mathrm{d}u}{\mathrm{d}t} = \frac{-\partial_t \phi}{1 + \partial_\vartheta \phi}\partial_\vartheta u_0^\phi - \frac{k}{(1 + \partial_\vartheta \phi)^2}\left(-\frac{\partial_{\vartheta\vartheta}\phi\, \partial_t \phi}{1 + \partial_\vartheta \phi} + \partial_\vartheta \partial_t \phi\right)\partial_k u_0^\phi$$

$$\frac{\mathrm{d}u}{\mathrm{d}\theta} = \frac{1}{1 + \partial_\vartheta \phi}\partial_\vartheta u_0^\phi - \frac{k\partial_{\vartheta\vartheta}\phi}{(1 + \partial_\vartheta \phi)^3}\partial_k u_0^\phi$$

$$\frac{\mathrm{d}^2 u}{\mathrm{d}\theta^2} = \left(\frac{1}{1 + \partial_\vartheta \phi}\frac{\mathrm{d}}{\mathrm{d}\vartheta} - \frac{k\partial_{\vartheta\vartheta}\phi}{(1 + \partial_\vartheta \phi)^3}\frac{\mathrm{d}}{\mathrm{d}k}\right)^2 u_0^\phi$$

and
$$\frac{dw}{dt} = \frac{\partial w}{\partial t} - \frac{\partial w}{\partial \vartheta} \frac{\partial_t \phi}{1 + \partial_\vartheta \phi}$$
$$\frac{dw}{d\theta} = \frac{1}{1 + \partial_\vartheta \phi} \frac{\partial w}{\partial \vartheta}$$
$$\frac{d^2 w}{d\theta^2} = \left(\frac{1}{1 + \partial_\vartheta \phi} \frac{d}{d\vartheta} \right)^2 w.$$

Thus, we get
$$(5.9) -\frac{\partial_t \phi}{1 + \partial_\vartheta \phi} \partial_\vartheta u_0^\phi + \frac{k}{(1 + \partial_\vartheta \phi)^2} \left(\partial_{\vartheta\vartheta} \phi \frac{\partial_t \phi}{1 + \partial_\vartheta \phi} - \partial_\vartheta \partial_t \phi \right) \partial_k u_0^\phi$$
$$+ \partial_t w - \frac{\partial_t \phi}{1 + \partial_\vartheta \phi} \partial_\vartheta w$$
$$= k^2 D \left(\left(\frac{1}{1 + \partial_\vartheta \phi} \frac{\partial}{\partial \vartheta} - \frac{k \partial_{\vartheta\vartheta}\phi}{(1 + \partial_\vartheta \phi)^3} \frac{\partial}{\partial k} \right)^2 u_0^\phi + \left(\frac{1}{1 + \partial_\vartheta \phi} \frac{\partial}{\partial \vartheta} \right)^2 w \right)$$
$$- \omega \frac{1}{1 + \partial_\vartheta \phi} \left(\partial_\vartheta u_0^\phi - \frac{k(\partial_{\vartheta\vartheta}\phi)}{(1 + \partial_\vartheta \phi)^2} \partial_k u_0^\phi + \partial_\vartheta w \right)$$
$$- (k^2 D \partial_{\vartheta\vartheta} u_0 - \omega \partial_\vartheta u_0 + f(u_0)) + f(u_0^\phi + w)$$

where we used (5.3) in the last equation.

Our next goal is to separate the critical modes, which involve the dynamics of ϕ, from the damped noncritical modes using the eigenfunctions of the linearization $\mathcal{L}$ given by
$$\mathcal{L} v = k^2 D \partial_{\vartheta\vartheta} v - \omega \partial_\vartheta v + f'(u_0(\vartheta; k)) v.$$
This will be accomplished using Bloch waves which we introduce next.

5.2. Bloch-wave analysis

We briefly recall from §4.1 some of the properties of the operator $\mathcal{L}$ as they serve as a motivation to introduce the Bloch-wave transform. In the notation of (4.8), the eigenfunctions $v(\vartheta)$ of the linearization $\mathcal{L}$ about $u_0(\vartheta; k)$ on $L^2(\mathbb{R})$ are given by
$$(5.10) \qquad v(\vartheta) = e^{-i\ell \vartheta / k} v(\vartheta; i\ell).$$
For our purposes, it is more convenient to parametrize solutions by the imaginary wave number ℓ, and we shall therefore use the notation
$$\check{v}(\vartheta, \ell) := v(\vartheta; i\ell)$$
throughout the rest of §5. As shown in §4.1, for each $\ell \in \mathbb{R}$, the functions $\check{v}(\cdot, \ell)$ are the 2π-periodic eigenfunctions of the operator $\check{\mathcal{L}}_\ell$ given by
$$\check{\mathcal{L}}_\ell \check{v} = k^2 D \left(\partial_\vartheta - \frac{i\ell}{k} \right)^2 \check{v} - \omega_0 \left(\partial_\vartheta - \frac{i\ell}{k} \right) \check{v} + f'(u_0(\vartheta)) \check{v}.$$

We observe that $\check{\mathcal{L}}_\ell$ coincides with the operator $\mathcal{L}_\nu$ for $\nu = i\ell$ discussed in §4.1. It is convenient in this section, however, to indicate explicitly when operators act on 2π-periodic functions which we shall do by using the superscript $\check{\ }$.

We also record that
$$(5.11) \qquad \check{v}(\vartheta, \ell + k) = e^{i\vartheta} \check{v}(\vartheta, \ell),$$

and we can therefore restrict ℓ to the interval $[-k/2, k/2)$. Furthermore, as in §4.1, we denote by $\check{v}_j(\vartheta, \ell)$ the eigenfunctions associated with the ordered branches $\lambda_j(i\ell)$ of eigenvalues of $\mathcal{L}_\ell$ for $j \in \mathbb{N}$. In particular, by Hypothesis 4.4, we have $\lambda_1 = \lambda_{\text{lin}}$.

We now turn to the Bloch-wave transform $\mathcal{J}$ which can be considered as a generalization of the Fourier transform $\mathcal{F}$. We will state only those properties of the Bloch-wave transform that we shall need subsequently and refer to [46, 51, 54] for their proofs and also for additional properties of the Bloch-wave transform. For every sufficiently smooth and rapidly enough decaying function w in ϑ-space, there are functions $\check{w}(\vartheta, \ell)$ that are 2π-periodic in ϑ and satisfy

$$\check{w}(\vartheta, \ell + k) = \mathrm{e}^{\mathrm{i}\vartheta} \check{w}(\vartheta, \ell) \tag{5.12}$$

with the property that w is represented via

$$w(\vartheta) = \int_{-k/2}^{k/2} \mathrm{e}^{-\mathrm{i}\ell\vartheta/k} \check{w}(\vartheta, \ell) \, \mathrm{d}\ell. \tag{5.13}$$

We shall write

$$\check{w} = \mathcal{J} w.$$

The rationale for the Bloch transform is as follows: If we denote the, slightly rescaled, Fourier transform of w by $\hat{w}$, then we have

$$
\begin{aligned}
w(\vartheta) &= \int_{-\infty}^{\infty} \mathrm{e}^{-\mathrm{i}\ell\vartheta/k} \hat{w}(\ell) \, \mathrm{d}\ell = \sum_{j \in \mathbb{Z}} \int_{-k/2}^{k/2} \mathrm{e}^{-\mathrm{i}\vartheta(\ell+jk)/k} \hat{w}(\ell + jk) \, \mathrm{d}\ell \\
&= \int_{-k/2}^{k/2} \mathrm{e}^{-\mathrm{i}\ell\vartheta/k} \left[\sum_{j \in \mathbb{Z}} \mathrm{e}^{-\mathrm{i}j\vartheta} \hat{w}(\ell + jk) \right] \mathrm{d}\ell =: \int_{-k/2}^{k/2} \mathrm{e}^{-\mathrm{i}\ell\vartheta/k} \check{w}(\vartheta, \ell) \, \mathrm{d}\ell,
\end{aligned}
$$

which is the desired Bloch-wave representation. Similar to the Fourier transform, the Bloch-wave transform can be defined for tempered distributions. We remark that the Bloch transform of the product of two functions w_1 and w_2 in ϑ-space is given by the convolution

$$\mathcal{J}[w_1 \cdot w_2](\vartheta, \ell) = [\check{w}_1 *_{\mathcal{J}} \check{w}_2](\vartheta, \ell) = \int_{-k/2}^{k/2} \check{w}_1(\vartheta, \ell - \tilde{\ell}) \check{w}_2(\vartheta, \tilde{\ell}) \, \mathrm{d}\tilde{\ell} \tag{5.14}$$

of their Bloch transforms $\check{w}_1$ and $\check{w}_2$ in Bloch space. Furthermore, if $w_1(\vartheta)$ is 2π-periodic in ϑ and the support of the Fourier transform $\hat{w}_2$ of a complex-valued function $w_2(\vartheta)$ lies in $(-1/2, 1/2)$, then we have

$$
\begin{aligned}
\mathcal{J}[w_1 w_2](\vartheta, \ell) &= \sum_{j \in \mathbb{Z}} \mathcal{F}[w_1 w_2](j + \ell) \mathrm{e}^{\mathrm{i}j\vartheta} = \sum_{j \in \mathbb{Z}} \mathrm{e}^{\mathrm{i}j\vartheta} \int_{-\frac{1}{2}}^{\frac{1}{2}} \hat{w}_1(j + \ell - \tilde{\ell}) \hat{w}_2(\tilde{\ell}) \, \mathrm{d}\tilde{\ell} \\
&= \sum_{j \in \mathbb{Z}} \mathrm{e}^{\mathrm{i}j\vartheta} \int_{-\frac{1}{2}}^{\frac{1}{2}} \hat{w}_1(j) \hat{w}_2(\tilde{\ell}) \delta_{\ell - \tilde{\ell}} \, \mathrm{d}\tilde{\ell} = \sum_{j \in \mathbb{Z}} \hat{w}_1(j) \mathrm{e}^{\mathrm{i}j\vartheta} \hat{w}_2(\ell) \\
&= w_1(\vartheta) \hat{w}_2(\ell)
\end{aligned}
$$

so that

$$\mathcal{J}[w_1 w_2](\vartheta, \ell) = w_1(\vartheta) \hat{w}_2(\ell). \tag{5.15}$$

The analytic properties of the Bloch-wave transform are based on a generalization of Parseval's identity

$$\int_{-\infty}^{\infty} |u(\vartheta)|^2 \, d\vartheta = 2\pi k \int_0^{2\pi} \int_{-k/2}^{k/2} |\check{u}(\vartheta,\ell)|^2 \, d\ell \, d\vartheta.$$

As a consequence, the Bloch-wave transform is an isomorphism from the space $H^m(n)$, equipped with the norm

$$\|u\|_{H^m(n)} = \|u\rho_{\rm w}^n\|_{H^m} \qquad \rho_{\rm w}(\vartheta) = \sqrt{1+\vartheta^2},$$

into the space $H^{m,n}_{\rm Bloch}$ of functions $\check{u}(\vartheta,\ell)$ that are 2π-periodic in ϑ, satisfy (5.12), and whose norm

$$\|\check{u}\|_{H^{m,n}_{\rm Bloch}} = \sum_{i=0}^{m}\sum_{j=0}^{n} \int_0^{2\pi} \int_{-k/2}^{k/2} |\partial_\vartheta^i \partial_\ell^j \check{u}(\vartheta,\ell)|^2 \, d\ell \, d\vartheta$$

is finite. As mentioned above, the proofs of the preceding statements can be found in [46, 51, 54].

Bloch-wave transform allows us to analyse differential operators with spatially-periodic coefficients. In Bloch space, such operators are multiplication operators. Since we are interested in functions without prescribed behaviour at infinity, i.e. in functions which do not necessarily decay to zero, we employ a method already used in [52] to extend multiplication operators from the space L^2 of square-integrable functions to the space $L^2_{\rm ul}$ of uniformly locally square-integrable functions equipped with the norm

$$\|u\|_{L^2_{\rm ul}} = \sup_{x\in\mathbb{R}} \int_x^{x+1} |u(y)|^2 \, dy.$$

We recall the definition

$$H^m_{\rm ul} = \left\{ u: \mathbb{R} \to \mathbb{R}; \, \|u\|_{H^m_{\rm ul}} = \|u\|_{H^m(x,x+1)} < \infty \text{ with } \lim_{y\to 0}\|u - T_y u\|_{H^m_{\rm ul}} \to 0 \right\}$$

where $[T_y u](x) := u(x+y)$.

LEMMA 5.3. *Let $m, s \in \mathbb{Z}$ with $m+s \geq 0$ and $m \geq 0$. Consider a function*

$$\check{\mathcal{M}}: \quad \mathbb{R} \longrightarrow L(H^{m+s}_{\rm per}(0,2\pi), H^m_{\rm per}(0,2\pi)), \qquad \ell \longmapsto \check{\mathcal{M}}(\ell)$$

which is at least $\mathcal{C}^2$ with respect to the Bloch wave number ℓ. The bounded linear operator $\mathcal{M}$ defined by

$$\mathcal{M}: \quad H^{m+s}(2) \longrightarrow H^m(2), \qquad u \longmapsto \mathcal{J}^{-1}(\check{\mathcal{M}}\mathcal{J}u)$$

can then be extended to a bounded linear operator

$$\mathcal{M}: \quad H^{m+s}_{\rm ul} \longrightarrow H^m_{\rm ul},$$

which we denote by the same letter, with norm

$$\|\mathcal{M}\|_{L(H^{m+s}_{\rm ul}, H^m_{\rm ul})} \leq C(m,s) \|\check{\mathcal{M}}\|_{\mathcal{C}^2_{\rm b}((-k/2,k/2), L(H^{m+s}_{\rm per}, H^m_{\rm per}))}$$

where $C(m,s)$ does not depend on $\mathcal{M}$.

PROOF. Choose a function $\chi \in \mathcal{C}_0^\infty$ so that its support is contained in $[-1,1]$ and $\sum_{j\in\mathbb{Z}} \chi(x+j) \equiv 1$. Next, pick $u \in H^{m+s}_{\rm ul}$ and set $u_j(x) = u(x)\chi(x-j)$. Since $u_j \in H^{m+s}(2)$, we know that $\mathcal{M}u_j \in H^m(2)$ on account of the results in [54, Lemma 5.4] (this is the crucial step which allows us now to extend the operator).

We define $(\mathcal{M}u)(x) := \sum_{j \in \mathbb{Z}} (\mathcal{M}u_j)(x)$: While this sum does not converge in H_{ul}^m, it converges locally to a function in H_{ul}^m with a norm bounded from above by

$$C(m,s)\|\check{\mathcal{M}}\|_{\mathcal{C}_{\mathrm{b}}^2((-k/2,k/2), L(H_{\mathrm{per}}^{m+s}, H_{\mathrm{per}}^m))} \|u\|_{H_{\mathrm{ul}}^{m+s}}$$

since $\mathcal{M}u_j$ is concentrated around $x = j$ and decays like $1/(1+|x-j|^2)$. □

REMARK 5.4. It is not difficult to see that this lemma can be extended to multilinear operators.

5.3. Mode filters, and separation into critical and noncritical modes

Our goal is to separate the dynamics of the eigenmodes $\check{v}_1(\vartheta, \ell)$ associated with the critical eigenvalues $\lambda_1(\mathrm{i}\ell) = \lambda_{\mathrm{lin}}(\mathrm{i}\ell)$ from the remaining damped noncritical modes. We shall use mode filters to obtain this splitting.

First, there exists a number ℓ_1 with $0 < \ell_1 \ll 1$ so that the eigenvalue $\lambda_1(\mathrm{i}\ell)$ of $\check{\mathcal{L}}_\ell$ is to the right of the rest of the spectrum for each ℓ with $|\ell| < \ell_1$. Therefore, there exists an $\check{\mathcal{L}}_\ell$-invariant projection

$$\check{Q}^{\mathrm{c}}(\ell) = \frac{1}{2\pi\mathrm{i}} \int_\Gamma [\lambda - \check{\mathcal{L}}_\ell]^{-1} \, \mathrm{d}\lambda$$

onto the space spanned by $\check{v}_1(\vartheta, \ell)$, where $\Gamma \subset \mathbb{C}$ is a small circle that surrounds $\lambda_1(\mathrm{i}\ell)$ counter-clockwise in the complex plane and does not intersect the rest of the spectrum of $\mathcal{L}_\ell$ for this fixed ℓ.

Next, we choose a decreasing $\mathcal{C}_0^\infty$-cutoff function $\chi : \mathbb{R} \to [0,1]$ with

(5.16) $$\chi(\ell) = \begin{cases} 1 & \text{for } |\ell| \leq 1 \\ 0 & \text{for } |\ell| \geq 2. \end{cases}$$

We can now define

$$\check{P}_{\mathrm{fs}}^{\mathrm{c}}(\ell) = \check{Q}^{\mathrm{c}}(\ell)\chi\left(\frac{4\ell}{\ell_1}\right), \qquad \check{P}_{\mathrm{fs}}^{\mathrm{s}}(\ell) := 1 - \check{Q}^{\mathrm{c}}(\ell)\chi\left(\frac{4\ell}{\ell_1}\right),$$

$$\check{P}_{\mathrm{mf}}^{\mathrm{c}}(\ell) = \check{Q}^{\mathrm{c}}(\ell)\chi\left(\frac{8\ell}{\ell_1}\right), \qquad \check{P}_{\mathrm{mf}}^{\mathrm{s}}(\ell) := 1 - \check{Q}^{\mathrm{c}}(\ell)\chi\left(\frac{8\ell}{\ell_1}\right)$$

and

$$\check{P}^{\mathrm{c}}(\ell) = \check{Q}^{\mathrm{c}}(\ell)\chi\left(\frac{2\ell}{\ell_1}\right), \qquad \check{P}^{\mathrm{s}}(\ell) := 1 - \check{Q}^{\mathrm{c}}(\ell)\chi\left(\frac{16\ell}{\ell_1}\right).$$

It is easy to check that these operators commute for each fixed ℓ and satisfy

(5.17) $(1 - \check{P}^{\mathrm{c}})\check{P}_{\mathrm{fs}}^{\mathrm{c}} = 0 = (1 - \check{P}_{\mathrm{fs}}^{\mathrm{c}})\check{P}_{\mathrm{mf}}^{\mathrm{c}}, \quad (1 - \check{P}^{\mathrm{s}})\check{P}_{\mathrm{fs}}^{\mathrm{s}} = 0 = (1 - \check{P}^{\mathrm{s}})\check{P}_{\mathrm{mf}}^{\mathrm{s}}$

$\check{P}_{\mathrm{fs}}^{\mathrm{c}} + \check{P}_{\mathrm{fs}}^{\mathrm{s}} = 1, \qquad \check{P}_{\mathrm{mf}}^{\mathrm{c}} + \check{P}_{\mathrm{mf}}^{\mathrm{s}} = 1.$

Lastly, we set

$$\check{\lambda}^{\mathrm{c}}(\ell) = \lambda_1(\mathrm{i}\ell)$$

and define scalar-valued operators $\check{p}_{\mathrm{fs}}^{\mathrm{c}}(\ell)$ and $\check{p}_{\mathrm{mf}}^{\mathrm{c}}(\ell)$ implicitly by

(5.18) $[\check{p}_{\mathrm{fs}}^{\mathrm{c}}(\ell)\check{u}]\check{v}_1(\cdot, \ell) = \check{P}_{\mathrm{fs}}^{\mathrm{c}}(\ell)\check{u}, \qquad [\check{p}_{\mathrm{mf}}^{\mathrm{c}}(\ell)\check{u}]\check{v}_1(\cdot, \ell) = \check{P}_{\mathrm{mf}}^{\mathrm{c}}(\ell)\check{u}$

for 2π-periodic functions $\check{u}(\vartheta)$. An application of Lemma 5.3 shows that each of the operators above extends to a bounded operator on H_{ul}^{m+s}. The resulting operators will be denoted by the same letter but with the superscript $\check{}$ being dropped.

The mode filters $p_{\mathrm{mf}}^{\mathrm{c}}$ and $P_{\mathrm{mf}}^{\mathrm{s}}$ obtained in this fashion will now be used to separate the critical and noncritical modes in (5.9) posed on $\mathbb{R}$. We will use the operators p_{fs}^{c} and P_{fs}^{s} to limit the Fourier support of the critical modes. First, we write (5.9), given by

$$
\begin{aligned}
&-\frac{\partial_t \phi}{1+\partial_\vartheta \phi}\partial_\vartheta u_0^\phi + \frac{k}{(1+\partial_\vartheta \phi)^2}\left(\partial_{\vartheta\vartheta}\phi\frac{\partial_t\phi}{1+\partial_\vartheta\phi} - \partial_\vartheta\partial_t\phi\right)\partial_k u_0^\phi \\
&+\partial_t w - \frac{\partial_t\phi}{1+\partial_\vartheta\phi}\partial_\vartheta w \\
(5.19)\quad =\;& k^2 D\left(\left(\frac{1}{1+\partial_\vartheta\phi}\frac{\partial}{\partial\vartheta} - \frac{k\partial_{\vartheta\vartheta}\phi}{(1+\partial_\vartheta\phi)^3}\frac{\partial}{\partial k}\right)^2 u_0^\phi + \left(\frac{1}{1+\partial_\vartheta\phi}\frac{\partial}{\partial\vartheta}\right)^2 w\right) \\
&-\omega\frac{1}{1+\partial_\vartheta\phi}\left(\partial_\vartheta u_0^\phi - \frac{k(\partial_{\vartheta\vartheta}\phi)}{(1+\partial_\vartheta\phi)^2}\partial_k u_0^\phi + \partial_\vartheta w\right) \\
&-(k^2 D\partial_{\vartheta\vartheta}u_0 - \omega\partial_\vartheta u_0 + f(u_0)) + f(u_0^\phi + w)
\end{aligned}
$$

as

$$(5.20)\qquad [-B_0 + B_1(\partial_\vartheta\phi, w)]\partial_t\phi + \partial_t w = -\tilde{\mathcal{L}}_0\partial_\vartheta\phi + \mathcal{L}w + G(\partial_\vartheta\phi, w),$$

where

$$
\begin{aligned}
B_0\partial_t\phi &= (\partial_\vartheta u_0 + k\partial_k u_0\partial_\vartheta)\partial_t\phi \\
\tilde{\mathcal{L}}_0\partial_\vartheta\phi &= \mathcal{L}(k\partial_\vartheta\phi\,\partial_k u_0) + k^2 D(2\partial_\vartheta\phi\,\partial_{\vartheta\vartheta}u_0 + \partial_{\vartheta\vartheta}\phi\,\partial_\vartheta u_0) - \omega\partial_\vartheta\phi\,\partial_\vartheta u_0 \\
(5.21)\qquad &= k\left[-\mathcal{L}(\partial_\vartheta\phi\,\partial_\nu v_1) + kD(2\partial_\vartheta\phi\,\partial_{\vartheta\vartheta}u_0 + \partial_{\vartheta\vartheta}\phi\,\partial_\vartheta u_0) - c_{\text{p}}\partial_\vartheta\phi\,\partial_\vartheta u_0\right] \\
B_1(\partial_\vartheta\phi, w)\partial_t\phi &= \left(\partial_\vartheta u_0 - \frac{\partial_\vartheta u_0^\phi}{1+\partial_\vartheta\phi}\right)\partial_t\phi \\
&\quad +k\left(\frac{\partial_{\vartheta\vartheta}\phi\,\partial_k u_0^\phi}{(1+\partial_\vartheta\phi)^3}\partial_t\phi + \left(\partial_k u_0 - \frac{\partial_k u_0^\phi}{(1+\partial_\vartheta\phi)^2}\right)\partial_\vartheta\partial_t\phi\right) \\
&\quad -\frac{\partial_\vartheta w}{1+\partial_\vartheta\phi}\partial_t\phi
\end{aligned}
$$

and G is comprised of the remaining terms. In the calculation above, we used that $\partial_\nu v_1 = -\partial_k u_0$, a fact we established in §4.2. Before continuing, we also remark that

$$
\begin{aligned}
B_1(\partial_\vartheta\phi, w) &= O(|\partial_\vartheta\phi| + |w|) \\
G(\partial_\vartheta\phi, w) &= O(|\partial_\vartheta\phi|^2 + |w|^2).
\end{aligned}
$$

Our goal is to replace (5.20) with the system

$$(5.22)\quad\begin{aligned}\partial_t P_{\text{fs}}^{\text{c}}B_0\phi &= P_{\text{fs}}^{\text{c}}\tilde{\mathcal{L}}_0\partial_\vartheta\phi + P_{\text{mf}}^{\text{c}}B_1(\partial_\vartheta\phi, w)\partial_t\phi - P_{\text{mf}}^{\text{c}}G(\partial_\vartheta\phi, w) \\
\partial_t w &= \mathcal{L}w + P_{\text{fs}}^{\text{s}}B_0\partial_t\phi - P_{\text{fs}}^{\text{s}}\tilde{\mathcal{L}}_0\partial_\vartheta\phi - P_{\text{mf}}^{\text{s}}B_1(\partial_\vartheta\phi, w)\partial_t\phi \\
&\quad + P_{\text{mf}}^{\text{s}}G(\partial_\vartheta\phi, w)\end{aligned}$$

for (ϕ, w). Subtracting the first from the second equation and using (5.17), we see that solutions of (5.22) give solutions of (5.20). Alternatively, we may consider the system

$$(5.23)\quad\begin{aligned}\partial_t p_{\text{fs}}^{\text{c}}B_0\phi &= p_{\text{fs}}^{\text{c}}\tilde{\mathcal{L}}_0\partial_\vartheta\phi + p_{\text{mf}}^{\text{c}}B_1(\partial_\vartheta\phi, w)\partial_t\phi - p_{\text{mf}}^{\text{c}}G(\partial_\vartheta\phi, w) \\
\partial_t w &= \mathcal{L}w + P_{\text{fs}}^{\text{s}}B_0\partial_t\phi - P_{\text{fs}}^{\text{s}}\tilde{\mathcal{L}}_0\partial_\vartheta\phi - P_{\text{mf}}^{\text{s}}B_1(\partial_\vartheta\phi, w)\partial_t\phi \\
&\quad + P_{\text{mf}}^{\text{s}}G(\partial_\vartheta\phi, w)\end{aligned}$$

for (ϕ, w), where the first equation is now scalar-valued. Inspecting (5.18) and exploiting that the eigenfunctions $\check{v}_1(\vartheta, \ell)$ satisfy a linear equation, we see that (5.22) and (5.23) are equivalent. To make (5.23) well-posed, we shall require that (ϕ, w) satisfy

$$\text{(5.24)} \quad \operatorname{supp} \mathcal{F}[\phi] \subset \mathcal{I} := \left\{ \ell;\, \chi\left(\frac{4\ell}{\ell_1}\right) = 1 \right\}$$

and

$$\text{(5.25)} \quad (1 - P^{\mathrm{s}})w = 0$$

for all $t \geq 0$. Since P^{s} commutes with $\mathcal{L}$, it follows from (5.17) and (5.23) that (5.25) is true for all $t > 0$ whenever it is met at $t = 0$.

It remains to check whether (5.24) is respected by (5.23) and to calculate the operator $p_{\mathrm{fs}}^{\mathrm{c}} B_0$ to see whether (5.23) is well posed. Due to the properties of the multiplier $p_{\mathrm{mf}}^{\mathrm{c}}$, we know that

$$\operatorname{supp} \mathcal{F}\left[p_{\mathrm{mf}}^{\mathrm{c}} (B_1(\partial_\vartheta \phi, w)\partial_t \phi - G(\partial_\vartheta \phi, w)) \right] \Subset \mathcal{I}$$

for any sufficiently smooth function ϕ. Next, we see from (5.21) that the operators B_0 and $\tilde{\mathcal{L}}_0$ have 2π-periodic coefficients in ϑ and are multipliers in Bloch space which allows us to use Lemma 5.3. For any function ϕ that satisfies (5.24), we then obtain

$$\begin{aligned}
\check{P}_{\mathrm{fs}}^{\mathrm{c}} \mathcal{J}[B_0 \phi] &= \check{P}_{\mathrm{fs}}^{\mathrm{c}}(\ell) \mathcal{J}[B_0 \phi](\vartheta, \ell) \\
&\stackrel{(5.15)}{=} \hat{\phi}(\ell) \chi\left(\frac{4\ell}{\ell_1}\right) \check{Q}^{\mathrm{c}}(\ell) \left(\partial_\vartheta u_0(\vartheta) + \mathrm{O}(\ell)\right) \\
&= \hat{\phi}(\ell) \chi\left(\frac{4\ell}{\ell_1}\right) (1 + \mathrm{O}(\ell)) \check{v}_1(\vartheta, \ell) \\
&\stackrel{(5.24)}{=} \left[(1 + \mathrm{O}(\ell_1)) \hat{\phi}(\ell) \right] \check{v}_1(\vartheta, \ell),
\end{aligned}$$

where the $\mathrm{O}(\ell_1)$ term is a multiplier. In particular, the term $[\ldots](\ell)$ has support in $\mathcal{I}$. Therefore, using the definition (5.18) of $p_{\mathrm{fs}}^{\mathrm{c}}$ and denoting the operator associated with the $\mathrm{O}(\ell_1)$ term by B_3, we get

$$\text{(5.26)} \quad p_{\mathrm{fs}}^{\mathrm{c}} B_0 \phi = (1 + B_3)\phi$$

for all ϕ that satisfy (5.24), where B_3 has norm $\|B_3\| = \mathrm{O}(\ell_1)$ and respects (5.24), i.e. $\operatorname{supp} \mathcal{F}[B_3 \phi] \subset \mathcal{I}$. Since similar arguments apply to the multiplier $\tilde{\mathcal{L}}_0$, we see that (5.24) is indeed preserved by (5.23) as claimed.

Next, for all (ϕ, w) for which $(\partial_\vartheta \phi, w)$ is small and ϕ satisfies (5.24), the first equation of (5.23) can be written as

$$\partial_t \phi = \left[1 + B_3 + p_{\mathrm{mf}}^{\mathrm{c}} B_1(\partial_\vartheta \phi, w) \right]^{-1} \left[p_{\mathrm{fs}}^{\mathrm{c}} \tilde{\mathcal{L}}_0 \partial_\vartheta \phi - p_{\mathrm{mf}}^{\mathrm{c}} G(\partial_\vartheta \phi, w) \right].$$

Substituting this expression for $\partial_t \phi$ into the second equation of (5.23) for w, we arrive at the system

$$\text{(5.27)} \quad \begin{aligned}
\partial_t \phi &= \left[1 + B_3 + p_{\mathrm{mf}}^{\mathrm{c}} B_1(\partial_\vartheta \phi, w) \right]^{-1} \left[p_{\mathrm{fs}}^{\mathrm{c}} \tilde{\mathcal{L}}_0 \partial_\vartheta \phi - p_{\mathrm{mf}}^{\mathrm{c}} G(\partial_\vartheta \phi, w) \right] \\
\partial_t w &= \mathcal{L} w - P_{\mathrm{fs}}^{\mathrm{s}} \tilde{\mathcal{L}}_0 \partial_\vartheta \phi + P_{\mathrm{mf}}^{\mathrm{s}} G(\partial_\vartheta \phi, w) \\
&\quad + \left[P_{\mathrm{fs}}^{\mathrm{s}} B_0 - P_{\mathrm{mf}}^{\mathrm{s}} B_1(\partial_\vartheta \phi, w) \right] \left[1 + B_3 + p_{\mathrm{mf}}^{\mathrm{c}} B_1(\partial_\vartheta \phi, w) \right]^{-1} \times \\
&\quad \times \left[p_{\mathrm{fs}}^{\mathrm{c}} \tilde{\mathcal{L}}_0 \partial_\vartheta \phi - p_{\mathrm{mf}}^{\mathrm{c}} G(\partial_\vartheta \phi, w) \right].
\end{aligned}$$

Thus, we accomplished an effective splitting of the critical modes ϕ and the non-critical modes w.

We now replace ϕ by $\psi = \partial_\vartheta \phi$ and obtain

$$\partial_t \psi = \partial_\vartheta \left[1 + B_3 + p^{\mathrm{c}}_{\mathrm{mf}} B_1(\psi, w)\right]^{-1} \left[p^{\mathrm{c}}_{\mathrm{fs}} \tilde{\mathcal{L}}_0 \psi - p^{\mathrm{c}}_{\mathrm{mf}} G(\psi, w)\right]$$

(5.28)
$$\partial_t w = \mathcal{L} w - P^{\mathrm{s}}_{\mathrm{fs}} \tilde{\mathcal{L}}_0 \psi + P^{\mathrm{s}}_{\mathrm{mf}} G(\psi, w)$$
$$+ \left[P^{\mathrm{s}}_{\mathrm{fs}} B_0 - P^{\mathrm{s}}_{\mathrm{mf}} B_1(\psi, w)\right] \left[1 + B_3 + p^{\mathrm{c}}_{\mathrm{mf}} B_1(\psi, w)\right]^{-1} \times$$
$$\times \left[p^{\mathrm{c}}_{\mathrm{fs}} \tilde{\mathcal{L}}_0 \psi - p^{\mathrm{c}}_{\mathrm{mf}} G(\psi, w)\right].$$

which we also write in short as

(5.29) $$\partial_t \mathcal{V} = \Lambda \mathcal{V} + \mathcal{N}(\mathcal{V}),$$

where $\mathcal{V} = (\psi, w)$, Λ is a linear operator, and $\mathcal{N}(\mathcal{V}) = \mathrm{O}(|\mathcal{V}|^2)$. We record that it is easy to check, using (4.21), (4.22) and (5.15), that the spectrum of the operator

$$\partial_\vartheta (1 + B_3)^{-1} p^{\mathrm{c}}_{\mathrm{mf}} \tilde{\mathcal{L}}_0$$

near $\lambda = 0$ is indeed given by the linear dispersion curve $\check{\lambda}^{\mathrm{c}}(\ell)$ with the associated eigenmodes given approximately by the Fourier modes $\exp(-i\ell\vartheta/k)$. Since the linear part of the system (5.28) is lower-triangular, we can find a bounded lower-triangular operator S that diagonalizes (5.29) so that

(5.30) $$S^{-1} \Lambda S = \mathrm{diag}(\lambda^{\mathrm{c}}, \Lambda^{\mathrm{s}}).$$

In particular, if we set $(v^{\mathrm{c}}, v^{\mathrm{s}}) := S^{-1} \mathcal{V}$, then we have $v^{\mathrm{c}} = \psi$ and $P^{\mathrm{s}} v^{\mathrm{s}} = v^{\mathrm{s}}$. In these coordinates, (5.29) becomes

(5.31)
$$\partial_t v^{\mathrm{c}} = \lambda^{\mathrm{c}} v^{\mathrm{c}} + \partial_\vartheta p^{\mathrm{c}}_{\mathrm{mf}} \mathcal{N}(v^{\mathrm{c}}, v^{\mathrm{s}})$$
$$\partial_t v^{\mathrm{s}} = \Lambda^{\mathrm{s}} v^{\mathrm{s}} + P^{\mathrm{s}}_{\mathrm{mf}} \mathcal{N}(v^{\mathrm{c}}, v^{\mathrm{s}}),$$

where $\mathcal{N}$ is a smooth nonlinear mapping from $H^{m+2}_{\mathrm{ul}} \times H^{m+2}_{\mathrm{ul}}$ into H^m_{ul} for every $m \geq 1$, and

$$\mathcal{N}^{\mathrm{c}}(v^{\mathrm{c}}, v^{\mathrm{s}}) := \partial_\vartheta p^{\mathrm{c}}_{\mathrm{mf}} \mathcal{N}(v^{\mathrm{c}}, v^{\mathrm{s}})$$

maps $H^{m+2}_{\mathrm{ul}} \times H^{m+2}_{\mathrm{ul}}$ into H^s_{ul} for each $s \geq 0$.

We emphasize that (5.31) therefore has the same properties as (3.28), whence we can follow the proofs in §3 almost line by line to finish the proof of the approximation result for reaction-diffusion equations.

5.4. Estimates for residuals and errors

We shall solve (5.31) for $(v^{\mathrm{c}}, v^{\mathrm{s}})$ in the space $\mathcal{X}_m \times \mathcal{X}_m$, where $\mathcal{X}_m := H^m_{\mathrm{ul}}$. We remark that v^{c} will, in fact, lie in H^s_{ul} for each $s \geq 0$.

LEMMA 5.5. *The operators λ^{c} and Λ^{s} are sectorial in $\mathcal{X}_m$ for every $m \geq 0$. Furthermore, there exist constants $C_0 > 0$ and $\sigma > 0$ such that the semigroups $\mathrm{e}^{\lambda^{\mathrm{c}} t}$ and $\mathrm{e}^{\Lambda^{\mathrm{s}} t}$ generated by these operators satisfy*

$$\|\mathrm{e}^{\lambda^{\mathrm{c}} t}\|_{\mathcal{X}_m \to \mathcal{X}_m} \leq C_0$$
$$\|\mathrm{e}^{\lambda^{\mathrm{c}} t} \partial_\vartheta\|_{\mathcal{X}_m \to \mathcal{X}_m} \leq C_0 t^{-1/2}$$
$$\|\mathrm{e}^{\Lambda^{\mathrm{s}} t}\|_{\mathcal{X}_m \to \mathcal{X}_m} \leq C_0 \mathrm{e}^{-\sigma t}$$

for all $t \geq 0$ and $m \geq 0$.

The second estimate in the preceding lemma provides decay of the semigroup for large $t \gg 1$ which we shall exploit below when we estimate the growth rate of solutions to (5.31).

PROOF. The operator Λ differs from the sectorial operator $D\partial_{\vartheta\vartheta}$ by a relatively bounded perturbation and is therefore also sectorial. Thus, by [22], Λ generates an analytic semigroup, and the growth rates of $e^{\Lambda t}$ are determined by the spectrum of Λ. In particular, $e^{\Lambda^s t}$ decays with some exponential rate. The singularity $t^{-1/2}$ for the center part is due to the parabolic profile of $\operatorname{Re} \lambda^c$ at $\ell = 0$ which allows us to apply Lemma 5.3 to $\mathcal{M}(\ell) = \delta\ell e^{\delta^{-2}\tilde{\lambda}^c(\delta\ell)T}$ with $T = \delta^2 t$. $\square$

We are now in a position to compute the evolution of residuals and errors. Upon substituting the ansatz
$$(v^c, v^s) = \left(\delta q\left(\delta((c_p - c_g)t - \vartheta/k), \delta^2 t\right), 0\right)$$
into (5.31) and computing the residual, we obtain
$$\operatorname{Res}_c(\delta q, 0) = \delta^3\left[-\partial_T q + \frac{\lambda_{\text{lin}}''(0)}{2}\partial_{XX}q + \frac{\omega_{\text{nl}}''(k)}{2}\partial_X(q^2)\right] + O(\delta^4)$$
$$\operatorname{Res}_s(\delta q, 0) = O(\delta^2).$$
Indeed, the second equation follows by using that the nonlinearity is quadratic, while the first equation follows from the calculation in §4.3. Thus, we see again that $q(X,T)$ should satisfy the Burgers equation
$$(5.32) \qquad \partial_T q = \frac{\lambda_{\text{lin}}''(0)}{2}\partial_{XX}q + \frac{\omega_{\text{nl}}''(k)}{2}\partial_X(q^2).$$

Thus, we fix integers $M \geq 1$ and $n \geq M + 3$, and pick a solution $q \in \mathcal{C}([0, T_0], H^n_{\text{ul}})$ of the Burgers equation (5.32). To derive error estimates, it is advantageous to add higher-order corrections to the above approximation.

LEMMA 5.6. *Fix positive integers n, m, M with $n \geq M + m + 3$, then there exists an improved approximation $(V^c, V^s) \in \mathcal{C}([0, T_0/\delta^2], H^n_{\text{ul}})$ such that the following is true. There exist $\delta_0 > 0$ and $C_{\text{res}} > 0$ such that*
$$\sup_{t \in [0, T_0/\delta^2]} \|V^c(\cdot, t; \delta) - q(\delta \cdot, \delta^2 t)\|_{\mathcal{X}_m} \leq C_{\text{res}}\delta$$
$$\sup_{t \in [0, T_0/\delta^2]} (\|V^c(\cdot, t; \delta)\|_{\mathcal{X}_m} + \|V^s(\cdot, t; \delta)\|_{\mathcal{X}_m}) \leq C_{\text{res}}$$
$$\sup_{t \in [0, T_0/\delta^2]} \|\operatorname{Res}_c(\delta V^c(\cdot, t; \delta), \delta^2 V^s(\cdot, t; \delta))\|_{\mathcal{X}_m} \leq C_{\text{res}}\delta^{M+3}$$
$$\sup_{t \in [0, T_0/\delta^2]} \|\operatorname{Res}_s(\delta V^c(\cdot, t; \delta), \delta^2 V^s(\cdot, t; \delta))\|_{\mathcal{X}_m} \leq C_{\text{res}}\delta^{M+2}$$
for all $\delta \in (0, \delta_0)$.

PROOF. Higher-order corrections $\Phi(X, T; \delta)$ and $u_1(\vartheta, X, T; \delta)$ in physical coordinates can be obtained as outlined in §4.3. Afterwards, we use the mode filters to transform these solutions into (v^c, v^s) form. While V^c has compact support in Fourier space, Φ and q do not. However the difference by the cut-off in Fourier space is $O(\delta^n)$ due to the concentration at the Bloch wave number $\ell = 0$ (see [54]). We identified Bloch space with Fourier space since v^c for fixed ℓ is one-dimensional. $\square$

Using the higher-order approximations, we introduce the scaled errors R^c and R^s by setting

$$\begin{aligned} v^c &= \delta V^c + \delta^{M+1} R^c \\ v^s &= \delta^2 V^s + \delta^{M+2} R^s. \end{aligned}$$

Substituting this ansatz into (3.28), and using the approximation properties of (V^c, V^s), we obtain exactly the same system as the one investigated in §3.9. Thus, following the analysis presented in §3.9, we obtain the following result which finishes the proof of Theorem 4.7.

PROPOSITION 5.7. *For each fixed positive integers n, m, M with $n \geq M+m+3$, there exists constants $\delta_1 > 0$ and $C_1 > 0$ such that we have*

$$\sup_{t \in [0, T_0/\delta^2]} \|R^c(t)\|_{\mathcal{X}_m} + \sup_{t \in [0, T_0/\delta^2]} \|R^s(t)\|_{\mathcal{X}_m} \leq C_1$$

for all $\delta \in (0, \delta_1)$.

5.5. Proofs of the theorems from §4.4

The proof of Theorem 4.6 follows closely that of Theorem 3.2 except that we adapt the proof of Theorem 4.7 instead of Theorem 3.5, and we therefore omit the details.

To prove Theorem 4.8, we observe that admissible solutions $q(X, T)$ lead to finite phase differences. In detail, we obtain

$$\lim_{\theta \to \infty} |\vartheta(\theta) - \vartheta_{\text{approx}}(\theta)| \leq C_1 \delta^m,$$

where $\vartheta(\theta)$ and $\vartheta_{\text{approx}}(\theta)$ are computed from the true solution $\phi(\vartheta, t)$ and the approximation $\delta V^c(\delta \vartheta, \vartheta^2 t)$. We can then reconstruct the phase from the wave number by integrating starting at $-\infty$.

Lastly, we comment on the proof of Theorem 4.9. From Proposition 5.7, one finds the estimate

$$|\vartheta(\theta) - \vartheta_{\text{approx}}(\theta + \phi_0(t))| \leq C\delta^{M+1}|\theta|$$

and therefore

$$|u(\theta, t) - u_{\text{approx}}(\theta, t)| \leq C_1 |\vartheta(\theta) - \vartheta_{\text{approx}}(\theta + \phi_0(t))| + C_1 |\partial_\vartheta \phi - V^c| \leq C\delta^{M+1}(|\theta| + 1).$$

Furthermore, we know that

$$\partial_t \phi_0(t) = \partial_t \phi(0, t) = O(\delta^2)$$

which yields the required estimate on the phase.

CHAPTER 6

Validity of the inviscid Burgers equation in reaction-diffusion systems

We discuss the evolution of modulated wave trains for wave-number modulations of small, but finite, size. The relevant ansatz in this situation is of the form

$$(6.1) \quad u_0(\omega_{\rm nl}(k_0)x - k_0 x - \delta^{-1}\Phi(X,T); k_0 + \partial_X \Phi(X,T)), \qquad (X,T) = (\delta x, \delta t).$$

This is the scaling considered by Howard and Kopell [23] who showed formally that $\Phi(X,T)$ ought to satisfy the inviscid phase equation

$$(6.2) \quad \partial_T \Phi + \omega_{\rm nl}(k_0 + \partial_X \Phi) - \omega_{\rm nl}(k_0) = 0,$$

while the wave number $q(X,T) = \partial_X \Phi(X,T)$ satisfies the hyperbolic conservation law

$$(6.3) \quad \partial_T q + \partial_X \omega_{\rm nl}(k_0 + q) = 0.$$

Since (6.3) is a conservation law, shocks will typically form in finite time. Due to the break-down of regularity during the formation of shocks, we can only expect to prove validity results over time intervals $[0, T_1/\delta]$, where $T_1 > 0$ is so small that the solution $q(X,T)$ has no shocks on $[0, T_1]$.

6.1. An illustration: The Ginzburg–Landau equation

To illustrate the concepts, we briefly review the set-up considered in [40] for the Ginzburg–Landau equation.

Our starting point is once more equation (3.13)

$$(6.4)\begin{aligned}\partial_t r &= \partial_{xx} r - 2r - \psi^2 - \psi^2 r - 2\alpha(\partial_x r)\psi - \alpha\partial_x \psi - \alpha(\partial_x \psi)r - 3r^2 - r^3 \\ \partial_t \psi &= \partial_{xx}\psi + \partial_x\left(\frac{\alpha\partial_{xx} r + 2(\partial_x r)\psi}{1+r} - \alpha\psi^2 - 2\beta r - \beta r^2\right)\end{aligned}$$

for amplitude and wave number corrections of the wave trains of the Ginzburg–Landau equation with $k = 0$.

In line with (6.1), we substitute the long-wave ansatz

$$(r,\psi)(x,t) = (W,q)(X,T), \qquad (X,T) = (\delta x, \delta t)$$

into (6.4), and get

$$\begin{aligned}\delta\partial_T W &= \delta^2 \partial_{XX} W - 2W - q^2 - q^2 W - 2\delta\alpha(\partial_X W)q - \delta\alpha\partial_X q \\ &\quad - \delta\alpha(\partial_X q)W - 3W^2 - W^3 \\ \partial_T q &= \delta\partial_{XX} q - \partial_X(2\beta W + \alpha q^2 + \beta W^2) + \partial_X\left(\frac{\delta^2 \alpha \partial_{XX} W + 2\delta\partial_X W}{1+W}\right).\end{aligned}$$

Choosing $0 < \delta \ll 1$ and neglecting terms that are formally of order $O(\delta)$, we obtain the system

$$0 = -2W - q^2 - q^2 W - 3W^2 - W^3$$
$$\partial_T q = -\partial_X(2\beta W + \alpha q^2 + \beta W^2).$$

For $|q| < 1$, the first equation is satisfied by $W = \sqrt{1-q^2} - 1$. Substituting this expression into the second equation, we see that the modulation $q(X,T)$ ought to satisfy the inviscid Burgers equation

$$\partial_T q + \partial_X\left((\alpha - \beta)q^2\right) = 0,$$

which we may also write as

(6.5) $$\partial_T q + \partial_X \omega_{\mathrm{nl}}(q) = 0,$$

where $\omega_{\mathrm{nl}}(k)$ denotes the nonlinear dispersion relation (3.3) of the Ginzburg–Landau equation. Local existence and uniqueness of solutions to the scalar first-order conservation law (6.5) are guaranteed by the method of characteristics or, for analytic initial data, by the Cauchy–Kowalevskaya theorem.

For various technical reasons, the validity results established in [40] are formulated for solutions $q(\delta x, \delta t)$ whose Fourier transform lives in the space

$$L_{\mathcal{F}}(\varrho, m) := \left\{ \hat{u} \in L^1(\mathbb{R}, \mathbb{C}); \ \|\hat{u}\| = \int_{\mathbb{R}} |\hat{u}(\ell)| e^{\varrho|\ell|}(1 + |\ell|^m)\,\mathrm{d}\ell < \infty \right\}$$

for $\varrho > 0$ and sufficiently large integers $m > 0$. For each $\hat{u}$ in $L(\varrho, m)$, the inverse Fourier transform u is analytic in a complex strip $\{z \in \mathbb{C}; \ |\mathrm{Im}\, z| < \varrho\}$ [28]. We shall use this and similar spaces in our analysis of reaction-diffusion systems.

6.2. Formal derivation of the conservation law

We repeat the formal derivation of the inviscid Burgers equation presented in [23, §2C] for reaction-diffusion systems which proceeds as in §4.3. Upon substituting the ansatz

(6.6) $$u(x,t) = u_0(\omega_{\mathrm{nl}}(k_0)t - k_0 x - \delta^{-1}\Phi(X,T); k_0 + \partial_X \Phi(X,T))$$

with $(X,T) = (\delta x, \delta t)$ into the reaction-diffusion system

$$\partial_t u = D\partial_{xx} u + f(u),$$

we obtain formally that

$$(\omega_{\mathrm{nl}}(k_0) - \partial_T \Phi)\partial_\theta u_0 = D(k_0 + \partial_X \Phi)^2 \partial_{\theta\theta} u_0 + f(u_0) + O(\delta).$$

Setting formally $\delta = 0$ and rearranging terms, we get

(6.7) $$D(k_0 + \partial_X \Phi)^2 \partial_{\theta\theta} u_0 - (\omega_{\mathrm{nl}}(k_0) - \partial_T \Phi)\partial_\theta u_0 + f(u_0) = 0,$$

where u_0 and its derivatives are evaluated as in (6.6). Formally treating (x,t) and (X,T) as independent variables, we find that the effective wave number k of the function $u_0(\cdot; k_0 + \partial_X \Phi)$ in (6.7) is equal to $k_0 + \partial_X \Phi$. Thus, comparing (6.7) with (4.5), we see that

$$\omega_{\mathrm{nl}}(k_0) - \partial_T \Phi = \omega_{\mathrm{nl}}(k_0 + \partial_X \Phi),$$

so that $\Phi(X,T)$ should indeed satisfy the inviscid phase equation

$$\partial_T \Phi + \omega_{\mathrm{nl}}(k_0 + \partial_X \Phi) - \omega_{\mathrm{nl}}(k_0) = 0.$$

Taking the derivative with respect to X, we see that the wave number $q = \partial_X \Phi$ should therefore be a solution of the inviscid Burgers equation

(6.8) $$\partial_T q + \partial_X \omega_{\mathrm{nl}}(k_0 + q) = 0.$$

6.3. Validity of the inviscid Burgers equation

Throughout this section, we assume that we are in the set-up introduced in §4.1 and §4.4.

THEOREM 6.1. *Assume that Hypothesis 4.1 is met. For any choice of $\varrho_0 > 0$ and integers $M \geq 1$ and $n \geq 3$, there are positive constants $\delta_1, \varepsilon_1, C_1, T_1$ such that the following is true. For each solution $q(\cdot, T) \in H^n_{\mathrm{ul}}$ of the conservation law (6.8) on the interval $[0, T_1]$ with*

$$\sup_{T \in [0, T_1]} \|q(\cdot, T)\|_{\mathcal{F}^{-1} L_{\mathcal{F}}(\varrho_0, 0)} \leq \varepsilon_1$$

and each $\delta \in (0, \delta_1)$, there are functions $(q_h, r_h)(\vartheta, t)$ in H^n_{ul} with

$$\sup_{t \in [0, T_1/\delta]} \|q_h(\cdot, t) - q(\delta \cdot, \delta t)\|_{H^n_{\mathrm{ul}}} \leq C_1 \delta$$

$$\sup_{t \in [0, T_1/\delta]} \|r_h(\cdot, t)\|_{H^n_{\mathrm{ul}}} \leq C_1 \sup_{T \in [0, T_1]} \|q(\cdot, T)\|^2_{H^n_{\mathrm{ul}}}$$

and a solution $u(\theta, t) = U(\vartheta, t)$ of the reaction-diffusion system (4.33) such that

$$\sup_{t \in [0, T_1/\delta]} \sup_{\vartheta \in \mathbb{R}} |U(\vartheta, t) - U_{\mathrm{approx}}(\vartheta, t)| \leq C_2 \delta^M,$$

where

$$U_{\mathrm{approx}}(\vartheta, t) = u_0(\vartheta; k(1 + q_h(\vartheta, t))) + r_h(\vartheta, t),$$

and ϑ and θ are related through (4.35) with

$$\phi(\vartheta, t) := \int_0^\vartheta q_h(\tilde{\vartheta}, t) \, d\tilde{\vartheta}.$$

The preceding theorem implies the following approximation result in the original variables (θ, t).

THEOREM 6.2. *Assume that Hypothesis 4.1 is met. For any choice of $\varrho_0 > 0$ and integers $M \geq 1$ and $n \geq 3$, there are positive constants ε_1, C_1, T_1 and δ_1 such that the following is true. For each solution $q(\cdot, T) \in H^n_{\mathrm{ul}}$ of the conservation law (6.8) on the interval $[0, T_1]$ with*

$$\sup_{T \in [0, T_1]} \|q(\cdot, T)\|_{\mathcal{F}^{-1} L_{\mathcal{F}}(\varrho_0, 0)} \leq \varepsilon_1$$

and each $\delta \in (0, \delta_1)$, there are functions $(q_h, r_h)(\vartheta, t)$ in H^n_{ul} with

$$\sup_{t \in [0, T_1/\delta]} \|q_h(\cdot, t) - q(\delta \cdot, \delta t)\|_{H^n_{\mathrm{ul}}} \leq C_1 \delta$$

$$\sup_{t \in [0, T_1/\delta]} \|r_h(\cdot, t)\|_{H^n_{\mathrm{ul}}} \leq C_1 \sup_{T \in [0, T_1]} \|q(\cdot, T)\|^2_{H^n_{\mathrm{ul}}},$$

a phase function $\phi_0(t)$ with

$$\sup_{t \in [0, T_1/\delta]} |\phi_0(t)| \leq \frac{1}{\delta} \sup_{T \in [0, T_1]} \|q(\cdot, T)\|^2_{H^n_{\mathrm{ul}}},$$

and a solution $u(\theta, t)$ of the reaction-diffusion system (4.33) such that
$$\sup_{t \in [0, T_1/\delta]} \sup_{\vartheta \in \mathbb{R}} |u(\theta, t) - u_{\text{approx}}(\theta, t)| \leq C_1 \delta^M,$$
where
$$u_{\text{approx}}(\theta, t) = u_0(\vartheta(\theta); k(1 + q_h(\vartheta(\theta), t))) + r_h(\vartheta(\theta), t),$$
and ϑ and θ are related through (4.35) with
$$\phi(\vartheta, t) := \phi_0(t) + \int_0^\vartheta q_h(\tilde{\vartheta}, t) \, \mathrm{d}\tilde{\vartheta}.$$

Thus, our results indicate that the solution profile is approximated accurately and, for the reasons outlined in Remark 3.13, we expect that the estimates for the profile are optimal for solutions with no additional properties. The position of the solution is given only up to an error of order $C_1 \|q\|^2/\delta$. We again believe that this is optimal.

The inviscid Burgers equation (6.8) can be written as
$$\partial_T q + \omega'_{\text{nl}}(k_0 + q)q_X = 0,$$
and we can think of its solutions $q(X, T)$ initially as waves that travel formally with speed $\omega'_{\text{nl}}(k_0 + q)$. In particular, the profile will, to leading order, move with the speed given by the group velocity c_g, and the estimate $\mathrm{O}(\|q\|^2/\delta)$ confirms this behaviour over time scales $\mathrm{O}(1/\delta)$. In this sense, Theorem 6.2 justifies that we named c_g the group velocity and interpreted it as the speed with which perturbations are transported along the wave train.

6.4. Proof of the theorems from §6.3

First, we note that Theorem 6.2 follows from Theorem 6.1 as in the proof of Theorem 4.9 in §5.5 except that $\partial_t \phi(0, t)$ is no longer of $\mathrm{O}(\delta^2)$: To get the correct estimate for $\phi(0, t)$, we first infer from (5.27) that
$$\partial_t \phi(0, t) = [1 + \mathrm{O}(|\ell_1|)] p^{\text{c}}_{\text{mf}} \left(\tilde{\mathcal{L}}_0 q(\delta \vartheta, \delta t) + \mathrm{O}(\delta + \|q\|^2) \right).$$
on account of the estimates obtained in Theorem 6.1. On the other hand, proceeding as in the derivation of (5.26), we obtain
$$p^{\text{c}}_{\text{mf}}(\tilde{\mathcal{L}}_0 q(\delta \vartheta, \delta t)) = \mathrm{O}(\delta).$$
Thus, $\partial_t \phi(0, t) = \mathrm{O}(\delta + \|q\|^2)$ as claimed and, consequently, $\sup_{t \in [0, T_1/\delta]} |\phi_0(t)| = \mathrm{O}(1 + \|q\|^2/\delta)$.

It therefore suffices to prove Theorem 6.1. As in the case of the Ginzburg–Landau equation [40], we restrict the class of admissible solutions of (6.8). For $\varrho > 0$, we set
$$L_{\mathcal{J}}(\varrho, m) :=$$
$$\left\{ \check{u} \in L^1([-k_0/2, k_0/2], H^m); \; \|\check{u}\|_{L_{\mathcal{J}}(\varrho, m)} := \int_{-k_0/2}^{k_0/2} \|\check{u}(\cdot, \ell)\|_{H^m} \mathrm{e}^{\varrho |\ell|} \, \mathrm{d}\ell < \infty \right\}$$
and remark that the 2π-periodic spatial variable x is not scaled in this space. Denoting the Bloch transform of a function u by $\check{u}$, we define the Banach space
$$\mathcal{X}_m^\varrho = \left\{ u : \mathbb{R} \to \mathbb{C}^d; \; \check{u} \in L_{\mathcal{J}}(\varrho, m) \right\}, \qquad \|u\|_{\mathcal{X}_m^\varrho} := \|\check{u}\|_{L_{\mathcal{J}}(\varrho, m)}.$$

Due to Sobolev embedding theorems, there is a constant $C(m) > 0$ for each $m \geq 1$ so that
$$\|\check{u} \star \check{v}\|_{L_{\mathcal{J}}(\varrho,m)} \leq C(m)\|\check{u}\|_{L_{\mathcal{J}}(\varrho,m)}\|\check{v}\|_{L_{\mathcal{J}}(\varrho,m)}$$
for any two functions $\check{u}, \check{v} \in L_{\mathcal{J}}(\varrho, m)$. Since $uv = \mathcal{J}^{-1}(\check{u} \star \check{v})$, we therefore obtain
$$\|uv\|_{\mathcal{X}_m^\varrho} \leq C(m)\|u\|_{\mathcal{X}_m^\varrho}\|v\|_{\mathcal{X}_m^\varrho}. \tag{6.9}$$
In particular, $\mathcal{X}_m^\varrho$ is an algebra under multiplication for $m \geq 1$. The constant $C(m)$ does not depend on $\varrho > 0$.

Fix $\varrho_0 > 0$ and integers $M \geq 1$ and $n \geq 3$. We also pick a solution $q(X,T)$ of the inviscid Burgers equation (6.8) in the space $\mathcal{F}^{-1}L_{\mathcal{F}}(\varrho_0, 0)$ on the time interval $[0, T_0]$. Scaling the independent variables via $(X,T) = (\delta(\vartheta - c_{\mathrm{p}}t), \delta t)$, we obtain that $q(\delta \cdot, T) \in \mathcal{X}_n^{\varrho_0/\delta}$ for all $T \in [0, T_0]$. As before, it is advantageous to add corrections to the approximation.

LEMMA 6.3. *For any $\varrho_1 \in (0, \varrho_0)$, there exist numbers $\delta_1 > 0$ and $C_{\mathrm{res}} > 0$ such that the following is true. For each $\delta \in (0, \delta_0)$, there are functions $(V^{\mathrm{c}}, V^{\mathrm{s}})$ such that*

$$\sup_{t \in [0, T_0/\delta]} \|V^{\mathrm{c}}(\cdot, t) - q(\delta\cdot, t)\|_{\mathcal{X}_m^{\varrho_1/\delta}} \leq C_{\mathrm{res}}\delta$$

$$\sup_{t \in [0, T_0/\delta]} \|V^{\mathrm{c}}(\cdot, t)\|_{\mathcal{X}_m^{\varrho_1/\delta}} \leq C_{\mathrm{res}}$$

$$\sup_{t \in [0, T_0/\delta]} \|V^{\mathrm{s}}(\cdot, t)\|_{\mathcal{X}_m^{\varrho_1/\delta}} \leq C_{\mathrm{res}} \sup_{T \in [0, T_0]} \|q(\cdot, T)\|_{H_{\mathrm{ul}}^n}^2$$

$$\sup_{t \in [0, T_0/\delta]} \|\mathrm{Res}_{\mathrm{c}}(V^{\mathrm{c}}(\cdot, t), V^{\mathrm{s}}(\cdot, t))\|_{\mathcal{X}_m^{\varrho_1/\delta}} \leq C_{\mathrm{res}}\delta^M$$

$$\sup_{t \in [0, T_0/\delta]} \|\mathrm{Res}_{\mathrm{s}}(V^{\mathrm{c}}(\cdot, t), V^{\mathrm{s}}(\cdot, t))\|_{\mathcal{X}_m^{\varrho_1/\delta}} \leq C_{\mathrm{res}}\delta^M.$$

PROOF. This follows as in Lemma 3.10 upon exploiting the diagonalization leading to (5.31). □

We introduce the critical noncritical parts $\delta^M R^{\mathrm{c}}$ and $\delta^M R^{\mathrm{s}}$ of the error by
$$v^{\mathrm{c}} = V^{\mathrm{c}} + \delta^M R^{\mathrm{c}}$$
$$v^{\mathrm{s}} = V^{\mathrm{s}} + \delta^M R^{\mathrm{s}}.$$

Substitution into (3.28) leads to
$$\partial_t R^{\mathrm{c}} = \lambda^{\mathrm{c}} R^{\mathrm{c}} + \rho g^{\mathrm{c}}(R^{\mathrm{c}}, R^{\mathrm{s}})$$
$$\partial_t R^{\mathrm{s}} = \Lambda^{\mathrm{s}} R^{\mathrm{s}} + g^{\mathrm{s}}(R^{\mathrm{c}}, R^{\mathrm{s}}).$$

For fixed constants D_{c} and D_{s}, there are constants so that the nonlinear terms satisfy
$$\|g^{\mathrm{c}}(R^{\mathrm{c}}, R^{\mathrm{s}})\|_{\mathcal{X}_m^\varrho} \leq C_{\mathrm{Res}} + C\|R^{\mathrm{c}}\|_{\mathcal{X}_m^\varrho} + C\|R^{\mathrm{s}}\|_{\mathcal{X}_m^\varrho} + \delta^M C(D_{\mathrm{c}}, D_{\mathrm{s}})$$
$$\|g^{\mathrm{s}}(R^{\mathrm{c}}, R^{\mathrm{s}})\|_{\mathcal{X}_{m-2}^\varrho} \leq C_{\mathrm{Res}} + C\|R^{\mathrm{c}}\|_{\mathcal{X}_m^\varrho} + C\|R^{\mathrm{s}}\|_{\mathcal{X}_m^\varrho} + \delta^M C(D_{\mathrm{c}}, D_{\mathrm{s}}),$$
uniformly in $\varrho \in [0, \varrho_1/\delta]$, provided
$$\|R^{\mathrm{c}}\|_{\mathcal{X}_m^\varrho} \leq D_{\mathrm{c}}, \qquad \|R^{\mathrm{s}}\|_{\mathcal{X}_m^\varrho} \leq D_{\mathrm{s}},$$
It becomes clear now that we cannot pursue the strategy used previously in §3.9 to prove that the errors are bounded, since this approach would require a factor $\delta^{1/2}$

in front of the estimates of g^c to work. Instead, we proceed as in [40], where the scale $\mathcal{X}_m^\varrho$ of Banach spaces has been used.

Pick a constant $K_0 > 0$. For each constant $K_1 > 0$, we may define a linear operator B via its symbol $\check{B}(\ell) = -K_1|\ell|$. We choose $K_1 \gg 1$ so large that the spectrum $\lambda_{K_1}(\ell)$ of $\lambda^c + B$ satisfies

$$\operatorname{Re} \lambda_{K_1}(\ell) \leq -K_0|\ell|$$

for the constant $K_0 > 0$ chosen above (thus, if the wave train is not sideband-unstable, any positive constant $K_1 > 0$ works). Next, we make the exponent ϱ in the family $\mathcal{X}_m^\varrho$ of Banach spaces smaller at a linear rate as time evolves by setting

(6.10) $$\varrho(t) := \frac{\varrho_1}{\delta} - K_1 t.$$

The requirement that $\varrho > 0$ therefore limits us to $0 \leq t \leq T_1/\delta$ for some $T_1 > 0$.

Next, we define the operator $\mathcal{S}(t)$ via its symbol $\check{\mathcal{S}}(t) = \mathrm{e}^{(\varrho_1/\delta - K_1 t)|\ell|}$ and introduce

$$\mathcal{R}^c(t) := \mathcal{S}(t) R^c(t), \qquad \mathcal{R}^s(t) := \mathcal{S}(t) R^s(t).$$

Note that $R^c(0) \in \mathcal{X}_m^{\varrho_1/\delta}$ is equivalent to $\mathcal{R}^c(0) \in \mathcal{X}_m^0$. The new error variables $\mathcal{R}^c$ and $\mathcal{R}^s$ satisfy

(6.11) $$\begin{aligned} \partial_t \mathcal{R}^c &= \lambda^c \mathcal{R}^c + B\mathcal{R}^c + \rho \mathcal{G}^c(\mathcal{R}^c, \mathcal{R}^s) \\ \partial_t \mathcal{R}^s &= \Lambda^s \mathcal{R}^s + B\mathcal{R}^s + \mathcal{G}^s(R^c, R^s). \end{aligned}$$

We shall work from now on in the space $\mathcal{X}_m := \mathcal{X}_m^0$ and denote its norm by $\|\cdot\|_m$. In this space, it has been shown in [40, §3.2] that the nonlinear terms obey the estimates

$$\begin{aligned} \|\mathcal{G}^c(\mathcal{R}^c, \mathcal{R}^s)\|_{\mathcal{X}_m} &\leq C_{\mathrm{res}} + C_q \|\mathcal{R}^c\|_{\mathcal{X}_m} + C_q \|\mathcal{R}^s\|_{\mathcal{X}_m} + \delta^M C(D_{\mathrm{c}}, D_{\mathrm{s}}) \\ \|\mathcal{G}^s(\mathcal{R}^c, \mathcal{R}^s)\|_{\mathcal{X}_{m-2}} &\leq C_{\mathrm{res}} + C_q \|\mathcal{R}^c\|_{\mathcal{X}_m} + C_q \|\mathcal{R}^s\|_{\mathcal{X}_m} + \delta^M C(D_{\mathrm{c}}, D_{\mathrm{s}}) \end{aligned}$$

for

$$\|\mathcal{R}^c\|_{\mathcal{X}_m} \leq D_{\mathrm{c}}, \qquad \|\mathcal{R}^s\|_{\mathcal{X}_m} \leq D_{\mathrm{s}}$$

and any fixed choice of positive constants D_{c} and D_{s}. Furthermore, we have that $C_q \to 0$ for $\|q\| \to 0$. The key is now the following optimal-regularity result proved in [40, §3.3].

LEMMA 6.4 ([40, §3.3]). *Fix $0 < \gamma < 1$, then there exists a constant $C_2 > 0$ with the following property: Pick any functions f^c and f^s with $f^c = p^c f^c$ and $f^s = P^s f^s$ for which $f^c(0)$ and $f^s(0)$ lie in the domains of $\lambda^c + B$ and $\Lambda^s + B$, respectively. The system*

$$\begin{aligned} \partial_t \mathcal{R}^c &= (\lambda^c + B)\mathcal{R}^c + \rho f^c, & \mathcal{R}^c(0) &= 0 \\ \partial_t \mathcal{R}^s &= (\Lambda^s + B)\mathcal{R}^s + f^s, & \mathcal{R}^c(0) &= 0 \end{aligned}$$

then has a unique solution on $[0, T_1/\delta]$, and

$$\begin{aligned} \|\mathcal{R}^c\|_{\mathcal{C}^{0,\gamma}([0,T_1/\delta],\mathcal{X}_m)} &\leq C_2 \|f^c\|_{\mathcal{C}^{0,\gamma}([0,T_1/\delta],\mathcal{X}_m)} \\ \|\mathcal{R}^s\|_{\mathcal{C}^{0,\gamma}([0,T_1/\delta],\mathcal{X}_m)} &\leq C_2 \|f^s\|_{\mathcal{C}^{0,\gamma}([0,T_1/\delta],\mathcal{X}_{m-2})}. \end{aligned}$$

The crucial assertion of the preceding lemma is, of course, the boundedness of solutions over the $O(1/\delta)$ times scale.

6.4. PROOF OF THE THEOREMS FROM §6.3

Using Lemma 6.4 together with property $C_q \to 0$ as $\|q\| \to 0$, we can now proceed as in §3.9 to prove that there are constants $C_3 > 0$ and $\delta_1 > 0$ so that

$$\sup_{t \in [0, T_1/\delta]} \|\mathcal{R}^c(t)\|_{\mathcal{X}_m} + \sup_{t \in [0, T_1/\delta]} \|\mathcal{R}^s(t)\|_{\mathcal{X}_m} \leq C_3,$$

uniformly in $\delta \in (0, \delta_1)$, for the solutions $\mathcal{R}^c$ and $\mathcal{R}^s$ of the full nonlinear problem (6.11). We omit the details.

CHAPTER 7

Modulations of wave trains near sideband instabilities

7.1. Introduction

In this section, we consider the dynamics of modulations of wave trains at the onset of sideband instabilities. A sideband instability is characterized by the condition that the second derivative $\lambda_{\text{lin}}''(0)$ of the linear dispersion relation changes sign as an appropriate systems parameter is varied. This sign change will lead to an instability of the wave train with respect to long-wavelength perturbations (i.e. perturbations with small wave number).

To be more precise, we denote the systems parameter by μ and assume that the wave train with wavenumber k exists for all μ close to zero, say. We expand the linear dispersion relation to get

$$(7.1) \quad \lambda_{\text{lin}}(\nu;\mu) = \lambda_{\text{lin}}'(0;\mu)\nu + \frac{1}{2}\lambda_{\text{lin}}''(0;\mu)\nu^2 + \frac{1}{6}\lambda_{\text{lin}}'''(0;\mu)\nu^3 + \frac{1}{24}\lambda_{\text{lin}}''''(0;\mu)\nu^4 + O(\nu^5)$$

where derivatives are taken with respect to ν. We are interested in the case where $\lambda_{\text{lin}}''(0;\mu)$ changes sign at $\mu = 0$. Thus, if $\lambda_{\text{lin}}''''(0;0) < 0$, which is a natural scenario in this context as it implies that the wave train is spectrally stable for $\mu < 0$, say, then the resulting instability for $\mu > 0$ is indeed induced by small wave numbers ℓ. In particular, sideband instabilities are modulational in nature, and we may therefore expect that they can be captured by adding appropriate corrections to the Burgers equation

$$(7.2) \quad 2\partial_T q = \lambda_{\text{lin}}''(0;\mu)\partial_{XX} q - \omega_{\text{nl}}''(0;\mu)\partial_X(q^2),$$

which itself becomes ill-posed once $\lambda_{\text{lin}}''(0;\mu) < 0$.

Our goal is to provide various validity results in this direction. The resulting equations depend on parameter scalings, and we therefore denote by $1/\delta$ the typical spatial length scale of modulations that we wish to capture.

Firstly, if we focus on the regime $|\mu| \leq C\delta^2$, then the Korteweg–de Vries equation (KdV) is the correct modulation equation that replaces the Burgers equation (7.2). Next, if we choose the scaling $\mu = \tilde{\mu}\delta$ with $\tilde{\mu} < 0$, then the resulting modulation equation is given by a dissipative Burgers-KdV equation.

A particularly interesting scenario arises if the third derivative $\lambda_{\text{lin}}'''(0;\mu)$ also vanishes at $\mu = 0$. This typically requires the adjustment of two parameters, or the presence of additional symmetries, and is therefore of codimension two. In this case, the Kuramoto–Sivashinsky equation takes the role of the Burgers equation in describing modulations of wave trains.

Lastly, we mention that there are other mechanisms that lead to the destabilization of wave trains. One such scenario are Hopf bifurcations where a second branch of the linear dispersion relation crosses the imaginary axis away from zero. In this case, we encounter a coupled system of two PDEs, namely the Burgers equation (describing wave-number modulations) and the complex Ginzburg–Landau equation (describing amplitude modulations of the Hopf modes) [17].

7.2. An illustration: The Ginzburg–Landau equation

Several different equations have been derived as phase equations for wave trains in the complex Ginzburg–Landau equation [2, 3, 20, 30, 39].

We concentrated in §3 on the wave train with $k = 0$. As can be read off (3.6), this wave train becomes sideband unstable at $1 + \alpha\beta = 0$. The linear dispersion relation $\lambda_{\text{lin}}(\nu)$ of the $k = 0$ wave train is quite degenerate: since the linearization is invariant under the reflection $x \mapsto -x$, all odd derivatives $\mathrm{d}^{2n+1}\lambda_{\text{lin}}/\mathrm{d}\nu^{2n+1}(0)$ vanish at $\nu = 0$. We also record from [1, (21)] that

$$(7.3) \qquad \lambda_{\text{lin}}''''(0) = -\frac{1}{2}\alpha^2(1 + \beta^2) < 0$$

for the wave train with $k = 0$. In particular, the governing equation near the sideband instability at $1 + \alpha\beta = 0$ is the Kuramoto–Sivashinsky equation. Its validity for phase modulations (but not for modulations of the wave number) has recently been proved in [2] near the sideband instability of the $k = 0$ wave train of the CGL. In §7, we shall give an approximation result for wave-number modulations in reaction-diffusion equations that should carry over to the CGL.

Sideband instabilities for wave trains with $k \neq 0$ yield the Korteweg–de Vries equation [20], at least when $\alpha \neq \beta$ (i.e. away from the real Ginzburg–Landau limit), since the third-order derivative of the linear dispersion relation at $\nu = 0$ does not vanish [1, (21)].

To illustrate how this higher-order PDEs arise as modulation equations, we shall derive the Kuramoto–Sivashinsky equation for the $k = 0$ wave train near its sideband instability which occurs when $1 + \alpha\beta = 0$. Starting point is, once again, equation (3.13)

$$(7.4)\begin{aligned}\partial_t r &= \partial_{xx} r - 2r - \psi^2 - \psi^2 r - 2\alpha(\partial_x r)\psi - \alpha\partial_x\psi - \alpha(\partial_x\psi)r - 3r^2 - r^3 \\ \partial_t \psi &= \partial_{xx}\psi + \partial_x\left(\frac{\alpha\partial_{xx} r + 2(\partial_x r)\psi}{1 + r} - \alpha\psi^2 - 2\beta r - \beta r^2\right)\end{aligned}$$

for amplitude and wave number corrections r and ψ of the $k = 0$ wave trains of the Ginzburg–Landau equation. We denote by δ the spatial length scale of the modulations we wish to capture. After picking an arbitrary constant $\kappa_2 \in \mathbb{R}$, we unfold the sideband instability in parameter space by setting

$$(7.5) \qquad 1 + \alpha\beta = \kappa_2 \delta^2.$$

Substituting the ansatz

$$r = \delta^6 W(\delta x, \delta^4 t), \qquad \psi = \delta^3 \Psi(\delta x, \delta^4 t)$$

into (7.4), and dividing the factors δ^6 and δ^7 in the equations for W and Ψ, respectively, we obtain

$$\begin{aligned}\delta^4\partial_T W &= \delta^2\partial_{XX}W - 2W - \Psi^2 - \delta^6 W\Psi^2 - 2\delta^4\alpha(\partial_X W)\Psi \\ &\quad - \alpha\delta^{-2}\partial_X\Psi - \delta^4\alpha(\partial_X\Psi)W - 3\delta^6 W^2 - \delta^{12}W^3 \\ \partial_T\Psi &= \delta^{-2}\partial_{XX}\Psi - \partial_X(2\beta W + \alpha\Psi^2) \\ &\quad + \delta^2\partial_X\left(\frac{\alpha\partial_{XX}W}{1+\delta^6 W} + \delta^2\frac{2(\partial_X W)\Psi}{1+\delta^6 W} - \delta^4\beta W^2\right)\end{aligned}$$

where $X = \delta x$ and $T = \delta^4 t$. Upon refining the leading-order solution $W = -\delta^{-2}\alpha\partial_X\Psi/2 + \mathrm{O}(1)$ to the first equation, we obtain that

$$W = -\frac{\alpha}{2\delta^2}\partial_X\Psi - \frac{\Psi^2}{2} - \frac{\alpha}{4}\partial_X^3\Psi + \mathrm{O}(\delta^2)$$

satisfies the first equation up to terms of order $\mathrm{O}(\delta^2)$. Substituting the expression for W into the equation for Ψ and using (7.5) gives

$$\begin{aligned}\partial_T\Psi &= \delta^{-2}(1+\alpha\beta)\partial_{XX}\Psi + (\beta-\alpha)\partial_X(\Psi^2) - \frac{\alpha(\alpha-\beta)}{2}\partial_X^4\Psi + \mathrm{O}(\delta^2) \\ &= \kappa_2\partial_{XX}\Psi + (\beta-\alpha)\partial_X(\Psi^2) - \frac{\alpha(\alpha-\beta)}{2}\partial_X^4\Psi + \mathrm{O}(\delta^2).\end{aligned}$$

Thus, setting $q := \Psi|_{\delta=0}$, we find that q ought to satisfy the Kuramoto–Sivashinsky equation

$$(7.6) \qquad \partial_T q = -\frac{\alpha(\alpha-\beta)}{2}\partial_X^4 q + \kappa_2\partial_{XX}q + (\beta-\alpha)\partial_X(q^2).$$

The factor in front of the fourth-order derivative coincides with (7.3) upon using that $1 + \alpha\beta = 0$.

As pointed out above, van Baalen considered sideband instabilities of the $k = 0$ wave train in [2]. He derived the Kuramoto–Sivashinsky equation

$$(7.7) \qquad \partial_T\Phi = -\frac{\alpha(\alpha-\beta)}{2}\partial_X^4\Phi + \kappa_2\partial_{XX}\Phi + (\beta-\alpha)(\partial_X\Phi)^2$$

for the phase Φ (and not the wave number q) and proved its validity in certain Sobolev spaces of spatially periodic functions under the technical assumption that $\alpha^2 < 1/2$.

7.3. Validity of the Korteweg–de Vries and the Kuramoto–Sivashinsky equation

We consider the reaction-diffusion system

$$(7.8) \qquad \partial_t u = D\partial_{xx}u + f(u;\mu)$$

with parameter $\mu \in \mathbb{R}^p$, and assume that the Hypotheses 4.1 and 4.4 from §4.1 are met at $\mu = 0$. In particular, the wave trains persist for all μ close to zero, and their linear dispersion relations are therefore given by

$$(7.9)\quad \lambda_{\mathrm{lin}}(\nu;\mu) = \lambda'_{\mathrm{lin}}(0;\mu)\nu + \frac{1}{2}\lambda''_{\mathrm{lin}}(0;\mu)\nu^2 + \frac{1}{6}\lambda'''_{\mathrm{lin}}(0;\mu)\nu^3 + \frac{1}{24}\lambda''''_{\mathrm{lin}}(0;\mu)\nu^4 + \mathrm{O}(\nu^5)$$

where the coefficients depend smoothly on μ, and derivatives are taken with respect to ν. Transforming into the θ-variables, we get

$$(7.10) \qquad \partial_t u = k^2 D\partial_{\theta\theta}u - \omega_{\mathrm{nl}}(k;\mu)\partial_\theta u + f(u;\mu).$$

We shall make frequent use of the notation introduced in §4.4. For the sake of clarity, we formulate the results in this section using only the ϑ-variables. We emphasize that all results transfer to the θ-variables in the same way as in §4.4.

We begin with the Kuramoto–Sivashinsky equation.

HYPOTHESIS 7.1. *We assume that* $\lambda_{\mathrm{lin}}''(0;0) = \lambda_{\mathrm{lin}}'''(0;0) = 0$ *and* $\lambda_{\mathrm{lin}}''''(0;0) < 0$.

We typically need to adjust a two-dimensional parameter $\mu \in \mathbb{R}^2$ to encounter the situation documented in Hypothesis 7.1.

THEOREM 7.2. *Assume that Hypotheses 4.1, 4.4 and 7.1 are met. Suppose that there is a smooth curve $\mu_*(\delta)$, defined for $0 \leq \delta \ll 1$ with $\mu_*(0) = 0$, such that*

$$(7.11) \quad \lambda_{\mathrm{lin}}''(0;\mu_*(\delta)) = \kappa_2 \delta^2 + \mathrm{O}(\delta^3), \qquad \lambda_{\mathrm{lin}}'''(0;\mu_*(\delta)) = \kappa_3 \delta + \mathrm{O}(\delta^2)$$

for appropriate constants $\kappa_j \in \mathbb{R}$. For each choice of integers $M \geq 1$ and $n \geq M+5$, and constants $C_0 > 0$ and $T_0 > 0$, there are constants $\delta_1 > 0$ and $C_1 > 0$ such that the following is true. For each $\delta \in (0, \delta_1)$ and each solution $q(X,T)$ of the Kuramoto–Sivashinsky equation

$$(7.12) \quad \partial_T q = \frac{1}{24}\lambda_{\mathrm{lin}}''''(0;0)\partial_X^4 q + \frac{1}{6}\kappa_3 \partial_X^3 q + \frac{1}{2}\kappa_2 \partial_{XX} q - \frac{1}{2}\omega_{\mathrm{nl}}''(k)\partial_X(q^2)$$

on $[0, T_0]$ with

$$\sup_{T \in [0,T_0]} \|q(\cdot,T)\|_{H^n_{\mathrm{ul}}} \leq C_0,$$

there are functions (q_h, r_h) with

$$\sup_{t \in [0,T_0/\delta^4]} \left\| q_h(\cdot,t) - q\left(\delta((c_{\mathrm{p}} - c_{\mathrm{g}})t - \vartheta/k), \delta^4 t\right) \right\|_{H^n_{\mathrm{ul}}} \leq C_1 \delta$$

$$\sup_{t \in [0,T_0/\delta^4]} \|r_h(\cdot,t)\|_{H^n_{\mathrm{ul}}} \leq C_1$$

and a solution $u(\theta,t) = U(\vartheta,t)$ of the reaction-diffusion system (7.10) for $\mu = \mu_*(\delta)$ such that

$$\sup_{t \in [0,T_0/\delta^4]} \sup_{\vartheta \in \mathbb{R}} |U(\vartheta,t) - U_{\mathrm{approx}}(\vartheta,t)| \leq C_1 \delta^{M+3},$$

where

$$U_{\mathrm{approx}}(\vartheta,t) = u_0(\vartheta; k(1 + \delta^3 q_h(\vartheta,t))) + \delta^6 r_h(\vartheta,t).$$

The phase function $\phi_0(t)$ needed for the formulation in the θ-variables satisfies $\sup_{t \in [0,T_0/\delta^4]} |\phi_0(t)| = \mathrm{O}(1)$.

Next, we consider validity of the dissipative Korteweg–de Vries equation.

HYPOTHESIS 7.3. *Assume that* $\lambda_{\mathrm{lin}}''(0;0) = 0$ *and* $\lambda_{\mathrm{lin}}'''(0;0) \neq 0$.

THEOREM 7.4. *Assume that Hypotheses 4.1, 4.4 and 7.3 are met. Suppose that there is a smooth curve $\mu_*(\delta)$, defined for $0 \leq \delta \ll 1$ with $\mu_*(0) = 0$, such that*

$$\lambda_{\mathrm{lin}}''(0;\mu_*(\delta)) = \kappa_2 \delta + \mathrm{O}(\delta^2)$$

for some positive constant $\kappa_2 > 0$. For each choice of integers $M \geq 1$ and $n \geq M+4$, and constants $C_0 > 0$ and $T_0 > 0$, there are constants $\delta_1 > 0$ and $C_1 > 0$ such that the following is true. For each $\delta \in (0,\delta_1)$ and each solution $q(X,T)$ of

$$(7.13) \quad \partial_T q = \frac{1}{6}\lambda_{\mathrm{lin}}'''(0;0)\partial_X^3 q + \frac{1}{2}\kappa_2 \partial_{XX} q - \frac{1}{2}\omega_{\mathrm{nl}}''(k)\partial_X(q^2)$$

on $[0, T_0]$ with
$$\sup_{T \in [0,T_0]} \|q(\cdot, T)\|_{H^n_{\mathrm{ul}}} \leq C_0,$$
there are functions (q_h, r_h) with
$$\sup_{t \in [0, T_0/\delta^3]} \left\|q_h(\cdot, t) - q\left(\delta((c_{\mathrm{p}} - c_{\mathrm{g}})t - \vartheta/k), \delta^3 t\right)\right\|_{H^n_{\mathrm{ul}}} \leq C_1 \delta$$
$$\sup_{t \in [0, T_0/\delta^3]} \|r_h(\cdot, t)\|_{H^n_{\mathrm{ul}}} \leq C_1$$
and a solution $u(\theta, t) = U(\vartheta, t)$ of the reaction-diffusion system (7.10) for $\mu = \mu_*(\delta)$ such that
$$\sup_{t \in [0, T_0/\delta^3]} \sup_{\vartheta \in \mathbb{R}} |U(\vartheta, t) - U_{\mathrm{approx}}(\vartheta, t)| \leq C_1 \delta^{M+2},$$
where
$$U_{\mathrm{approx}}(\vartheta, t) = u_0(\vartheta; k(1 + \delta^2 q_h(\vartheta, t))) + \delta^4 r_h(\vartheta, t).$$
The phase function $\phi_0(t)$ satisfies $\sup_{t \in [0, T_0/\delta^3]} |\phi_0(t)| = \mathrm{O}(1)$.

The result that we shall prove for the conservative Korteweg–de Vries equation is less satisfactory. We cannot exploit that the linear dispersion relation is dissipative at $\ell = 0$ since dissipativeness becomes noticeable only over time scales of length δ^{-4}. On the other hand, solutions to the Korteweg–de Vries equation exist for all times so that we should not run into the restrictions that we encountered and discussed in §6. However, we shall not exploit these properties, and our result below is therefore weaker and can probably be improved considerably.

THEOREM 7.5. *Assume that Hypotheses 4.1, 4.4 and 7.3 are met. Suppose also that $\mu_*(\delta)$ is a smooth curve, defined for $0 \leq \delta \ll 1$ with $\mu_*(0) = 0$, such that*

(7.14) $$\lambda''_{\mathrm{lin}}(0; \mu_*(\delta)) = \mathrm{O}(\delta^2).$$

For any choice of $\varrho_0 > 0$ and integers $M \geq 1$ and $n \geq 3$, there are positive constants $\delta_1, \varepsilon_1, C_1, T_1$ such that the following is true. For each $\delta \in (0, \delta_1)$ and each solution $q(X, T)$ of the Korteweg–de Vries equation

(7.15) $$\partial_T q = \frac{1}{6} \lambda'''_{\mathrm{lin}}(0;0) \partial_X^3 q - \frac{1}{2} \omega''_{\mathrm{nl}}(k) \partial_X (q^2)$$

on $[0, T_1]$ with
$$\sup_{T \in [0,T_1]} \|q(\cdot, T)\|_{\mathcal{X}^{\varrho_0}_n} \leq \varepsilon_1,$$
there are functions (q_h, r_h) with
$$\sup_{t \in [0, T_1/\delta^3]} \left\|q_h(\cdot, t) - q\left(\delta((c_{\mathrm{p}} - c_{\mathrm{g}})t - \vartheta/k), \delta^3 t\right)\right\|_{H^n_{\mathrm{ul}}} \leq C_1 \delta$$
$$\sup_{t \in [0, T_1/\delta^3]} \|r_h(\cdot, t)\|_{H^n_{\mathrm{ul}}} \leq C_1$$
and a solution $u(\theta, t) = U(\vartheta, t)$ of the reaction-diffusion system (7.10) for $\mu = \mu_*(\delta)$ such that
$$\sup_{t \in [0, T_1/\delta^3]} \sup_{\vartheta \in \mathbb{R}} |U(\vartheta, t) - U_{\mathrm{approx}}(\vartheta, t)| \leq C_1 \delta^{M+2},$$
where
$$U_{\mathrm{approx}}(\vartheta, t) = u_0(\vartheta; k(1 + \delta^2 q_h(\vartheta, t))) + \delta^4 r_h(\vartheta, t).$$
The phase function $\phi_0(t)$ satisfies $\sup_{t \in [0, T_1/\delta^3]} |\phi_0(t)| = \mathrm{O}(1)$.

We shall prove Theorems 7.2 and 7.5 in §7.4 and 7.5, respectively. We omitted the proof of Theorem 7.4 since it is similar to the one given for Theorem 7.2 except for the different scalings.

7.4. Proof of Theorem 7.2

We proceed in exactly the same way as in §3 and §5.4. Our starting point is again the system (5.31)

(7.16)
$$\begin{aligned}\partial_t v^c &= \lambda^c v^c - p_{\mathrm{mf}}^c \partial_\vartheta \mathcal{N}(v^c, v^s) \\ \partial_t v^s &= \Lambda^s v^s - P_{\mathrm{mf}}^s \mathcal{N}(v^c, v^s).\end{aligned}$$

The operators λ^c and Λ^s generate semigroups with properties analogous to those established in §5.4.

LEMMA 7.6. *Both λ^c and Λ^s are sectorial in $\mathcal{X}_m$, respectively. Moreover, for each integer $m \geq 0$, there are constants $C_0 > 0$ and $\sigma > 0$ such that the semigroups $\mathrm{e}^{\lambda^c t}$ and $\mathrm{e}^{\Lambda^s t}$ satisfy*

$$\begin{aligned}\|\mathrm{e}^{\lambda^c t}\|_{\mathcal{X}_m \to \mathcal{X}_m} &\leq C_0 \\ \|\mathrm{e}^{\lambda^c t} \partial_\vartheta\|_{\mathcal{X}_m \to \mathcal{X}_m} &\leq \frac{C_0}{t^{1/4}} \\ \|\mathrm{e}^{\Lambda^s t}\|_{\mathcal{X}_m \to \mathcal{X}_m} &\leq C_0 \mathrm{e}^{-\sigma t}\end{aligned}$$

for all $t \geq 0$.

PROOF. The first and third inequality follow as in Lemma 3.9. The reason that the second inequality is valid is due to the fact that λ^c is, by construction, the multiplication operator in Bloch space associated with the linear dispersion relation. More precisely, Hypothesis 7.1 implies that $\mathrm{Re}\,\lambda^c(\ell) \approx -\ell^4$. The factor $t^{-1/4}$ is now a consequence of Lemma 3.6 applied to the function

$$\check{\mathcal{M}}(\ell) = \delta \ell \mathrm{e}^{\delta^{-4} \check{\lambda}^c(\delta\ell) T}$$

with $T = \delta^4 t$. □

Starting with a solution $q(X, T)$ of the Kuramoto–Sivashinsky equation (7.12), we substitute the ansatz

$$(v^c, v^s) = \left(\delta^3 q\left(\delta((c_\mathrm{p} - c_\mathrm{g})t - \vartheta/k), \delta^4 t\right), 0\right)$$

into (7.16) and obtain the residuals

$$\begin{aligned}\mathrm{Res}_\mathrm{c}(\delta^3 q, 0) &= \delta^7 \left(-\partial_T q + \frac{1}{2}\kappa_2 \partial_{XX} q + \frac{1}{6}\kappa_3 \partial_X^3 q \right. \\ &\quad \left. + \frac{1}{24}\check{\lambda}^{c''''}(0)\partial_X^4 q - \frac{1}{2}\omega_{\mathrm{nl}}''(k)\partial_X(q^2)\right) + \mathrm{O}(\delta^8) \\ &= \mathrm{O}(\delta^8) \\ \mathrm{Res}_\mathrm{s}(\delta^3 q, 0) &= -\check{P}^s(\ell)\check{\mathcal{N}}(\delta^3 q)(\ell) = \mathrm{O}(\delta^6)\end{aligned}$$

where we used (7.11). Next, we record that the formal procedures outlined in §3.8 and §4.3 can again be used to provide approximations with smaller residuals.

LEMMA 7.7. *With m, n, M chosen as in Theorem 7.2, there are positive constants $\delta_1 > 0$ and $C_{\text{res}} > 0$ such that the following is true. For each $\delta \in (0, \delta_1)$, there exist functions (V^c, V^s) such that*

$$\sup_{t \in [0, T_0/\delta^4]} \|V^c(\cdot, t) - q(\delta \cdot, t)\|_{\mathcal{X}_m} \leq C_{\text{res}} \delta$$

$$\sup_{t \in [0, T_0/\delta^4]} \|V^s(\cdot, t)\|_{\mathcal{X}_m} \leq C_{\text{res}}$$

$$\sup_{t \in [0, T_0/\delta^4]} \|\text{Res}_c(\delta^3 V^c(\cdot, t), \delta^6 V^s(\cdot, t))\|_{\mathcal{X}_m} \leq C_{\text{res}} \delta^{M+7}$$

$$\sup_{t \in [0, T_0/\delta^4]} \|\text{Res}_s(\delta^3 V^c(\cdot, t), \delta^6 V^s(\cdot, t))\|_{\mathcal{X}_m} \leq C_{\text{res}} \delta^{M+6}$$

uniformly in δ.

We define the scaled errors R^c and R^s relative to the approximations obtained in the preceding lemma via

$$v^c = \delta^3 V^c + \delta^{M+3} R^c$$
$$v^s = \delta^6 V^s + \delta^{M+6} R^s.$$

Substitution into (7.16) gives the system

$$\partial_t R^c = \lambda^c R^c + \partial_\vartheta g^c(R^c, R^s)$$
$$\partial_t R^s = \Lambda^s R^s + g^s(R^c, R^s).$$

There is a constant C_0 such that

(7.17) $\|g^c(R^c, R^s)\|_{\mathcal{X}_m} \leq \delta^3 C_0 \|R^c\|_{\mathcal{X}_m} + \delta^4 C_{\text{Res}} + \delta^6 C_0 \|R^s\|_{\mathcal{X}_m}$
$\qquad\qquad\qquad\quad + \delta^{M+3} C(D_c, D_s)$

$\|g^s(R^c, R^s)\|_{\mathcal{X}_{m-2}} \leq C_{\text{Res}} + C_0 \|R^c\|_{\mathcal{X}_m} + \delta^3 C_0 \|R^s\|_{\mathcal{X}_m}$
$\qquad\qquad\qquad\quad + \delta^M C(D_c, D_s)$

where

$$\|R^c\|_{\mathcal{X}_m} \leq D_c, \qquad \|R^s\|_{\mathcal{X}_m} \leq D_s$$

for arbitrary, but fixed, constants D_c and D_s.

The rest of the proof proceeds as in §3.9. Using Lemma 7.6 and Gronwall's lemma 3.12 over the time scale $T = \delta^4 t$ gains the crucial factor δ which, taken together with the factor δ^3 on the right-hand side of (7.18), shows that the scaled errors stay bounded.

With regard to the θ-variables, we record that $\partial_t \phi(0, t) = \mathrm{O}(\delta^4)$ so that the phase shift $\sup_{t \in [0, T_0/\delta^4]} |\phi(0, t)| = \mathrm{O}(1)$ is bounded uniformly in δ as claimed.

7.5. Proof of Theorem 7.5

We proceed as in §6 to which we refer for the notation we use below. First, we pick $\varrho_0 > 0$, integers $M \geq 1$ and $n \geq M + 4$, and a solution $q(X, T)$ of the KdV equation (7.15) in $\mathcal{X}_n^\varrho$. Next, we substitute the resulting ansatz

$$(v^c, v^s) = \left(\delta^2 q\left(\delta((c_p - c_g)t - \vartheta/k), \delta^3 t\right), 0\right)$$

into (7.16) to obtain the residuals

$$\begin{aligned}
\mathrm{Res}_{\mathrm{c}}(\delta^2 q, 0) &= \delta^5\left(-\partial_T q + \frac{1}{6}\check{\lambda}^{c\prime\prime\prime}(0)\partial_X^3 q - \frac{1}{2}\omega_{\mathrm{nl}}''(k)\partial_X(q^2)\right) + \mathrm{O}(\delta^6) = \mathrm{O}(\delta^6) \\
\mathrm{Res}_{\mathrm{s}}(\delta^2 q, 0) &= -\check{P}^{\mathrm{s}}(\ell)\check{\mathcal{N}}(\delta^3 q)(\ell) = \mathrm{O}(\delta^4),
\end{aligned}$$

where we used (7.14). As before, we can construct approximations with smaller residuals.

LEMMA 7.8. *With ϱ_0 and M, n chosen as above, there are positive constants $\delta_1 > 0$ and $C_{\mathrm{res}} > 0$ such that the following is true. For each $\delta \in (0, \delta_1)$, there exist functions $(V^{\mathrm{c}}, V^{\mathrm{s}})$ such that*

$$\sup_{t\in[0,T_0/\delta^4]} \|V^{\mathrm{c}}(\cdot, t) - q(\delta\cdot, t)\|_{\mathcal{X}_m^\varrho} \le C_{\mathrm{res}}\delta$$

$$\sup_{t\in[0,T_0/\delta^4]} \|V^{\mathrm{s}}(\cdot, t)\|_{\mathcal{X}_m^\varrho} \le C_{\mathrm{res}}$$

$$\sup_{t\in[0,T_0/\delta^4]} \|\mathrm{Res}_{\mathrm{c}}(\delta^3 V^{\mathrm{c}}(\cdot, t), \delta^6 V^{\mathrm{s}}(\cdot, t))\|_{\mathcal{X}_m^\varrho} \le C_{\mathrm{res}}\delta^{M+2}$$

$$\sup_{t\in[0,T_0/\delta^4]} \|\mathrm{Res}_{\mathrm{s}}(\delta^3 V^{\mathrm{c}}(\cdot, t), \delta^6 V^{\mathrm{s}}(\cdot, t))\|_{\mathcal{X}_m^\varrho} \le C_{\mathrm{res}}\delta^{M+4},$$

where $\varrho = \varrho_0/\delta$.

The errors R^{c} and R^{s}, defined via

$$\begin{aligned}
v^{\mathrm{c}} &= \delta^2 V^{\mathrm{c}} + \delta^{M+2} R^{\mathrm{c}} \\
v^{\mathrm{s}} &= \delta^4 V^{\mathrm{s}} + \delta^{M+4} R^{\mathrm{s}},
\end{aligned}$$

satisfy the system

$$\begin{aligned}
\partial_t R^{\mathrm{c}} &= \lambda^{\mathrm{c}} R^{\mathrm{c}} + \partial_\vartheta g^{\mathrm{c}}(R^{\mathrm{c}}, R^{\mathrm{s}}) \\
\partial_t R^{\mathrm{s}} &= \Lambda^{\mathrm{s}} R^{\mathrm{s}} + g^{\mathrm{s}}(R^{\mathrm{c}}, R^{\mathrm{s}})
\end{aligned}$$

where, for an appropriate positive constant $C_0 > 0$,

$$\|g^{\mathrm{c}}(R^{\mathrm{c}}, R^{\mathrm{s}})\|_{\mathcal{X}_m^\varrho} \le \delta^2 C_{\mathrm{Res}} + \delta^2 C_0 \|R^{\mathrm{c}}\|_{\mathcal{X}_m^\varrho} + \delta^4 C_0 \|R^{\mathrm{s}}\|_{\mathcal{X}_m^\varrho} + \delta^{M+2} C(D_{\mathrm{c}}, D_{\mathrm{s}})$$
$$\|g^{\mathrm{s}}(R^{\mathrm{c}}, R^{\mathrm{s}})\|_{\mathcal{X}_{m-2}^\varrho} \le C_{\mathrm{Res}} + C_0 \|R^{\mathrm{c}}\|_{\mathcal{X}_m^\varrho} + \delta^2 C_0 \|R^{\mathrm{s}}\|_{\mathcal{X}_m^\varrho} + \delta^M C(D_{\mathrm{c}}, D_{\mathrm{s}})$$

uniformly in $\varrho \in [0, \varrho_0/\delta]$ for

$$\|R^{\mathrm{c}}\|_{\mathcal{X}_m^\varrho} \le D_{\mathrm{c}}, \qquad \|R^{\mathrm{s}}\|_{\mathcal{X}_m^\varrho} \le D_{\mathrm{s}}$$

where D_{c} and D_{s} are arbitrary but fixed.

As in §6, we need to exploit the scale of Banach spaces given by $\mathcal{X}_m^\varrho$. We begin by picking a constant $K_0 > 0$. For any given constant $K_1 > 0$, we may define the linear operator B via its symbol $\check{B}(\ell) = -K_1 \delta^2 |\ell|$. As in §6, we choose $K_1 \gg 1$ so large that the spectrum $\lambda_{K_1}(\ell)$ of $\lambda^{\mathrm{c}} + B$ satisfies

$$\mathrm{Re}\,\lambda_{K_1}(\ell) \le -K_0 \delta^2 |\ell|$$

for the constant $K_0 > 0$ chosen above: note that such a choice of K_1 is possible due to (7.14 and Hypotheses 4.4 and 7.3. Next, we define the operator $\mathcal{S}(t)$ via its symbol $\check{\mathcal{S}}(t) = \mathrm{e}^{(\varrho_0/\delta - K_1 \delta^2 t)|\ell|}$ and introduce

$$\mathcal{R}^{\mathrm{c}}(t) := \mathcal{S}(t) R^{\mathrm{c}}(t), \qquad \mathcal{R}^{\mathrm{s}}(t) := \mathcal{S}(t) R^{\mathrm{s}}(t)$$

7.5. PROOF OF THEOREM 7.5

which both live in $\mathcal{X}_m := \mathcal{X}_m^0$ with $\|\cdot\|_m$. The rescaled errors $\mathcal{R}^c$ and $\mathcal{R}^s$ satisfy

$$\partial_t \mathcal{R}^c = (\lambda^c + B)\mathcal{R}^c + \partial_\vartheta \mathcal{G}^c(\mathcal{R}^c, \mathcal{R}^s)$$
$$\partial_t \mathcal{R}^s = (\Lambda^s + B)\mathcal{R}^s + \mathcal{G}^s(R^c, R^s)$$

where

$$\|\mathcal{G}^c(\mathcal{R}^c, \mathcal{R}^s)\|_{\mathcal{X}_m} \leq \delta^2 C_{\text{Res}} + \delta^2 C_q \|\mathcal{R}^c\|_{\mathcal{X}_m} + \delta^4 C_q \|\mathcal{R}^s\|_{\mathcal{X}_m} + \delta^{M+2} C(D_c, D_s)$$
$$\|\mathcal{G}^s(\mathcal{R}^c, \mathcal{R}^s)\|_{\mathcal{X}_{m-2}} \leq C_{\text{Res}} + C_q \|\mathcal{R}^c\|_{\mathcal{X}_m} + \delta^2 C_q \|\mathcal{R}^s\|_{\mathcal{X}_m} + \delta^M C(D_c, D_s)$$

for

$$\|\mathcal{R}^c\|_{\mathcal{X}_m} \leq D_c, \qquad \|\mathcal{R}^s\|_{\mathcal{X}_m} \leq D_s$$

for positive constants C_q with $C_q \to 0$ as $\|q\| \to 0$. The rest of the proof follows exactly as in §6, and we therefore omit it.

CHAPTER 8

Existence and stability of weak shocks

We prove Theorem 4.10 and 4.12 by introducing an appropriate spatial dynamics formulation to which we apply the Kirchgässner reduction. The ideas behind this approach go back to Kirchgässner [29] and were later extended by Mielke and coworkers [18, 41].

8.1. Proof of Theorem 4.10

Suppose that $u(x,t) = u_*(x - c_* t, \omega_* t)$ satisfies (4.1) and (4.39), then

$$(u, v)(\xi, \tau) = (u_*, \partial_\xi u_*)(\xi, \tau)$$

has period 2π in τ and satisfies the modulated-wave equation

(8.1)
$$\begin{aligned}\partial_\xi u &= v \\ \partial_\xi v &= -D^{-1}[-\omega_* \partial_\tau u + c_* v + f(u)]\end{aligned}$$

where $\xi = x - c_* t$ and $\tau = \omega_* t$. We consider (8.1) on the function space $\mathcal{X} = H^1_{\mathrm{per}}(0, 2\pi) \times H^{1/2}_{\mathrm{per}}(0, 2\pi)$ and write $\mathbf{u} = (u, v) \in \mathcal{X}$. Equation (8.1) is equivariant with respect to the time shift $\mathcal{S}(\rho)$ defined by $[\mathcal{S}(\rho)\mathbf{u}](\tau) = \mathbf{u}(\tau + \rho)$ for $\rho \in [0, 2\pi]$. In other words, if $\mathbf{u}(\xi) \in \mathcal{X}$ is a solution of (8.1), so is $\mathcal{S}(\rho)\mathbf{u}(\xi)$ for each ρ.

Instead of investigating (8.1) for arbitrary c_* and ω_*, we will fix a primary wave number k_0 and concentrate on finding viscous shocks that have speed $c_{\mathrm{g}}(k_0) =: c_{\mathrm{g}}^0$. Since $c_{\mathrm{g}}(k) = \omega'_{\mathrm{nl}}(k)$ and due to our assumption that $\omega''_{\mathrm{nl}}(k_0) \neq 0$, we know that, for each given number c close to $c_{\mathrm{g}}(k_0)$, there exists a wave number k close to k_0 such that $c = c_{\mathrm{g}}(k)$. Repeating the proof given below for different wave numbers k close to k_0 gives shocks with an arbitrary speed c close to c_{g}^0. Thus, it suffices to prove Theorem 4.10 for the fixed speed c_{g}^0.

Therefore, from now on, we will set $c_* = c_{\mathrm{g}}^0$ in (8.1). From (4.44), we conclude that the frequency of shocks with speed c_{g}^0 and asymptotic wave numbers $k_\pm = k_0$ is $\omega_*^0 = k_0(c_{\mathrm{p}}^0 - c_{\mathrm{g}}^0)$. If we choose different wave numbers for $k_\pm$, the temporal frequency of shocks will vary as well. We therefore write $\omega_* = \omega_*^0 + \bar{\omega}$ so that $\bar{\omega}$ varies near zero.

We shall assume that $\omega_*^0 \neq 0$: If $\omega_*^0 = 0$, then we fix $\omega_* = 0$ and allow c_* to vary near c_{g}^0. The associated weak shocks are travelling waves which satisfy (8.1) with $\omega_* = 0$, i.e. an ODE. The primary wave train appears as a saddle-node periodic orbit in this ODE. Unfolding the vector field on the two-dimensional center manifold using the speed c_* gives the desired weak shocks as heteroclinic orbits. We shall omit the details since the analysis is similar to (but far easier than) the forthcoming analysis of the case $\omega_*^0 \neq 0$.

Using the definitions introduced above, (8.1) becomes

(8.2) $$\begin{aligned}\partial_\xi u &= v \\ \partial_\xi v &= -D^{-1}[-(\omega_*^0+\bar\omega)\partial_\tau u + c_g^0 v + f(u)]\end{aligned}$$

where $(u,v) \in \mathcal{X}$. The wave train with wave number k_0,

$$\begin{aligned} u &= u_0(\omega_0 t - k_0(\xi + c_g^0 t)) = u_0(k_0(c_p^0 - c_g^0)t - k_0\xi) = u_0(\tau - k_0\xi), \\ \omega_0 &= \omega_{\mathrm{nl}}(k_0)\end{aligned}$$

provides a solution to (8.2) with $\bar\omega = 0$. If we transform (8.2) according to

$$\mathbf{u} \longmapsto \mathbf{u}_0 + \mathbf{u}$$

where

(8.3) $$\mathbf{u}_0(\xi) = \begin{pmatrix} u_0(\cdot - k_0\xi) \\ -k_0\partial_\theta u_0(\cdot - k_0\xi) \end{pmatrix}$$

we obtain

(8.4) $$\begin{aligned}\partial_\xi u &= v \\ \partial_\xi v &= -D^{-1}[-(\omega_*^0+\bar\omega)\partial_\tau u + c_g^0 v + f'(u_0(\cdot - k_0\xi))u \\ &\qquad + g(u; u_0(\cdot - k_0\xi)) - \bar\omega\partial_\tau u_0(\cdot - k_0\xi)]\end{aligned}$$

where

$$g(u; u_0) := f(u_0+u) - f(u_0) - f'(u_0)u = \frac{1}{2}f''(u_0)[u,u] + \mathrm{O}(\|u\|^3).$$

We write this equation (8.4) abstractly as

(8.5) $$\partial_\xi \mathbf{u} = \mathcal{B}_*(\xi)\mathbf{u} + \bar\omega\mathcal{N}(\mathbf{u}+\mathbf{u}_0) + \mathcal{G}(\mathbf{u},\xi)$$

where

$$\begin{aligned}\mathcal{B}_*(\xi) &= \begin{pmatrix} 0 & 1 \\ -D^{-1}[-\omega_*^0\partial_\tau + f'(u_0(\cdot-k_0\xi))] & -D^{-1}c_g^0 \end{pmatrix} \\ \mathcal{N} &= \begin{pmatrix} 0 & 0 \\ D^{-1}\partial_\tau & 0 \end{pmatrix}, \qquad \mathcal{G}(\mathbf{u},\xi) = \begin{pmatrix} 0 \\ -D^{-1}g(u; u_0(\cdot-k_0\xi)) \end{pmatrix}.\end{aligned}$$

Alternatively, we can consider (8.4) in the temporally comoving frame $\sigma = \tau - k_0\xi$ which gives

(8.6) $$\begin{aligned}\partial_\xi u &= k_0\partial_\sigma u + v \\ \partial_\xi v &= k_0\partial_\sigma v - D^{-1}[-\omega_*^0\partial_\sigma u + c_g^0 v + f'(u_0(\cdot))u \\ &\qquad + g(u; u_0(\cdot)) - \bar\omega\partial_\sigma(u + u_0(\cdot))]\end{aligned}$$

or, in abstract form,

(8.7) $$\partial_\xi \mathbf{u} = [k_0\mathcal{T} + \mathcal{B}_*^0]\mathbf{u} + \bar\omega\mathcal{N}(\mathbf{u}+\mathbf{u}_0) + \mathcal{G}^0(\mathbf{u})$$

where $\mathcal{T} = \mathrm{diag}(\partial_\sigma, \partial_\sigma)$ generates the temporal shift and

$$\begin{aligned}\mathcal{B}_*^0 &= \begin{pmatrix} 0 & 1 \\ -D^{-1}[-\omega_*^0\partial_\sigma + f'(u_0(\cdot))] & -D^{-1}c_g^0 \end{pmatrix} \\ \mathcal{G}^0(\mathbf{u}) &= \begin{pmatrix} 0 \\ -D^{-1}g(u; u_0(\cdot)) \end{pmatrix}.\end{aligned}$$

Solutions to (8.5) and (8.7) are conjugated by the shift generated by $k_0\mathcal{T}$. We also consider the linearized equations

(8.8) $$\partial_\xi \mathbf{u} = \mathcal{B}_*(\xi)\mathbf{u}$$

and

(8.9) $$\partial_\xi \mathbf{u} = \left[k_0\mathcal{T} + \mathcal{B}_*^0\right]\mathbf{u}.$$

The coefficient matrix of (8.8) is periodic in ξ, while the coefficient matrices of (8.9) do not depend on ξ. From [41] we conclude that the Floquet exponents of (8.8) form a discrete set in the complex plane and that there exists a strongly continuous family of center projections $\mathcal{P}^c(\xi)$ that are $2\pi/k_0$-periodic in ξ and have finite-dimensional range. Thus, on account of [41, Theorem 3.4], the nonlinear equation (8.5) admits a nonautonomous center manifold $\mathcal{E}(\xi) = \mathcal{E}(\xi + 2\pi/k_0)$ that is tangent to $\mathrm{Rg}(\mathcal{P}^c(\xi))$ at $(\mathbf{u},\bar{\omega}) = 0$ and that contains all small bounded solutions to (8.5). Conjugation with the shift evolution $\mathcal{S}(k_0\xi)$ associated with $\partial_\xi \mathbf{u} = k_0\mathcal{T}\mathbf{u}$ gives an invariant center-manifold to (8.7) whose center subspace consists of the generalized eigenspace to the center eigenvalues $\nu \in i\mathbb{R}$ of $k_0\mathcal{T} + \mathcal{B}_*^0$. We remark that there does not seem to exist a center-manifold theorem for the equations of the type (8.7) where a hyperbolic structure $\mathcal{T}$ is mixed with a pseudo-elliptic structure $\mathcal{B}_*^0$. Still, smooth center-manifolds of (8.7) exist since they can be obtained from smooth center-manifolds of (8.5). We prefer to work with (8.7), since we avoid complications that are caused by the ambiguity in the definition of Floquet exponents.

We need one additional property of the center manifold. Recall that (8.2) is invariant under the action of the time-shift $\mathcal{S}(\rho)$. In a neighbourhood of the relative equilibrium $u_0(\tau - k_0\xi)$, this action correspondence to the following equivariance of the nonautonomous equation (8.5): if $\mathbf{u}(\xi;\xi_0)$ is a solution, so is $\mathbf{u}(\xi+\rho;\xi_0+\rho)$. In the construction of a center manifold, $\mathcal{G}(\mathbf{u},\xi)$ is replaced by

$$\mathcal{G}_{\mathrm{mod}}(\mathbf{u},\xi) := \chi(\|\mathbf{u}\|_\mathcal{X}^2)\mathcal{G}(\mathbf{u},\xi)$$

for some smooth cut-off function $\chi(r)$ that satisfies $\chi(r) = 1$ for $r < \delta \ll 1$ and $\chi(r) = 0$ for $r > 2\delta$. Since the norm of the Hilbert space $\mathcal{X}$ is invariant under the time shift $\mathcal{S}$, equivariance is preserved under the cut-off procedure. In particular, the flow on the center manifold commutes with the (affine) action of the circle group. We will use this fact extensively.

In the remaining part of the proof, we compute the generalized center eigenspace of $k_0\mathcal{T} + \mathcal{B}_*^0$ and use the result to calculate the expansion of the reduced vector field on the center manifold. We begin with the computation of the center eigenspace.

We consider the eigenvalue problem

$$\nu\mathbf{u} = \left[k_0\mathcal{T} + \mathcal{B}_*^0\right]\mathbf{u}$$

or, more explicitly,

$$\begin{aligned}\nu u &= k_0\partial_\sigma u + v \\ \nu v &= k_0\partial_\sigma v - D^{-1}[-\omega_*^0\partial_\sigma u + f'(u_0(\sigma))u + c_{\mathrm{g}}^0 v].\end{aligned}$$

This boundary-value problem has a solution in $\mathcal{X}$ if, and only if,

(8.10) $$k_0^2 D\left(\partial_\sigma - \frac{\nu}{k_0}\right)^2 u - \omega_0\left(\partial_\sigma - \frac{\nu}{k_0}\right)u + f'(u_0(\sigma))u = (c_{\mathrm{p}}^0 - c_{\mathrm{g}}^0)\nu u$$

has a nontrivial 2π-periodic solution. A comparison with the operator $\mathcal{L}_\nu$, defined in (4.9), shows that nontrivial solutions to (8.10) exist precisely when $\lambda = (c_{\mathrm{p}}^0 - c_{\mathrm{g}}^0)\nu$

is an eigenvalue of $\mathcal{L}_\nu$ for some $\nu \in i\mathbb{R}$. This, however, was excluded for $\lambda \neq 0$ and $\nu \neq 0$ in the nonresonance assumption of Hypothesis 4.3. Therefore, the only possible center eigenvalue occurs at $\nu = 0$. The same argument shows that the null space of $k_0 \mathcal{T} + \mathcal{B}_*^0$ is one-dimensional and spanned by $(\partial_\sigma u_0, -k_0 \partial_{\sigma\sigma} u_0)$. It remains to calculate generalized eigenvectors which are the solutions of the derivative of (8.10) with respect to ν evaluated in $u = \partial_\sigma u_0$. The point is that the eigenvalue problem

$$(8.11) \qquad k_0^2 D \left(\partial_\theta - \frac{\nu}{k_0} \right)^2 v - \omega_0 \left(\partial_\theta - \frac{\nu}{k_0} \right) v + f'(u_0(\theta)) v = \lambda_{\text{lin}}(\nu) v$$

for the operator $\mathcal{L}_\nu$, see (4.9), coincides to quadratic order in ν with (8.10) since

$$\lambda_{\text{lin}}(\nu) = (c_{\text{p}}^0 - c_{\text{g}}^0)\nu + O(\nu^2)$$

as calculated in (4.23). Thus, recalling the results from §4.2, we therefore see that the generalized eigenvector $\mathbf{u}_1$ exists. Its u-component is given as the unique (up to elements of the null space) solution to

$$\mathcal{L}_0 u_1 = 2 k_0 D \partial_{\sigma\sigma} u_0 - c_{\text{g}}^0 \partial_\sigma u_0,$$

while the v-component of the generalized eigenvector is given by

$$v_1 = \partial_\sigma u_0 - k_0 \partial_\sigma u_1.$$

Thus, comparing the equation for u_1 with (4.18), we conclude that $u_1 = -\partial_k u_0$ and therefore

$$\mathbf{u}_1 = \begin{pmatrix} -\partial_k u_0 \\ k_0 \partial_{k\sigma} u_0 + \partial_\sigma u_0 \end{pmatrix}.$$

Since $\lambda_{\text{lin}}''(0) \neq 0$ by assumption, we also see that the eigenvalue $\nu = 0$ has algebraic multiplicity equal to two.

We emphasize that the higher-dimensional eigenspace is generated precisely by our choice of the coordinate frame—for other choices of the speed c, the generalized center eigenspace would be one-dimensional and spanned by the translated wave trains. Thus, the group velocity can be interpreted as the unique speed for which the linearization about the wave train, computed in the frame moving with that speed, develops algebraic multiplicity two.

For the computation of the projection onto the generalized eigenspace, we will also need the generalized eigenspace of the adjoint $[k_0 \mathcal{T} + \mathcal{B}_*^0]_{\text{ad}}$, where we compute the adjoint with respect to the simpler L^2-scalar product instead of the scalar product in $\mathcal{X}$. Also, as we shall see below, we only need a basis vector, denoted by $\mathbf{u}_{\text{ad}}$, of the null space of the adjoint operator. The function $\mathbf{u}_{\text{ad}}$ satisfies the equation

$$\begin{aligned} -k_0 \partial_\sigma u - D^{-1} \omega_*^0 \partial_\sigma v - f'(u_0)^T D^{-1} v &= 0 \\ -k_0 \partial_\sigma v + u - D^{-1} c_{\text{g}}^0 v &= 0. \end{aligned}$$

If we set $\tilde{v} = D^{-1} v$, we obtain

$$k_0^2 D \partial_{\sigma\sigma} \tilde{v} + \omega_0 \partial_\sigma \tilde{v} + f'(u_0)^T \tilde{v} = 0$$

whose solution $\tilde{v} = u_{\text{ad}}$ we computed in §4.2. The null space of the L^2-adjoint of $k_0 \mathcal{T} + \mathcal{B}_*^0$ is therefore spanned by

$$\mathbf{u}_{\text{ad}} = \begin{pmatrix} k_0 D \partial_\sigma u_{\text{ad}} + c_{\text{g}}^0 u_{\text{ad}} \\ D u_{\text{ad}} \end{pmatrix}.$$

The next step is to calculate the vector field on the center manifold of (8.7). We parametrize the two-dimensional center manifold by $(\theta, \kappa) \in S^1 \times (-\delta, \delta)$ via
$$\mathbf{u} = -\kappa \mathbf{u}_1(\cdot - \theta) + \mathbf{u}_h(\theta, \kappa),$$
where $\mathbf{u}_h$ maps $\mathbb{R}^2$ into the orthogonal complement of the generalized eigenspace, belonging to the eigenvalue zero, of the adjoint $[k_0 \mathcal{T} + \mathcal{B}_*^0]_{\mathrm{ad}}$, and by simultaneously replacing σ by $\sigma - \theta$ in (8.7). In particular, the time-shift symmetry acts according to $\theta \mapsto \theta + \rho$. Since, as discussed above, the vector field on the center manifold respects this symmetry, it cannot depend on θ. To leading order, we therefore find

(8.12) $\qquad \partial_\xi \theta = \kappa + \mathrm{O}(|\bar{\omega}| + |\kappa|^2)$

(8.13) $\qquad \partial_\xi \kappa = \beta_\omega \bar{\omega} + \beta_2 \kappa^2 + \mathrm{O}(|\bar{\omega}|^2 + |\bar{\omega}\kappa| + |\kappa|^3).$

To compute the coefficients β_ω and β_2, we substitute the global parameterization

(8.14) $\qquad \mathbf{u} = \mathbf{u}_0(\cdot - \theta) - \kappa \mathbf{u}_1(\cdot - \theta) + \mathrm{O}(\theta^2 + \kappa^2)$

of the center manifold into the original equation (8.2), projecting $\mathcal{N}$ and $\mathcal{G}$ onto the center manifold using the spectral projection, and comparing the terms. We obtain

$$\beta_\omega = -\frac{\langle \mathbf{u}_{\mathrm{ad}}, \mathcal{N} \mathbf{u}_0 \rangle}{\langle \mathbf{u}_{\mathrm{ad}}, \mathbf{u}_1 \rangle} = -\frac{\langle D u_{\mathrm{ad}}, D^{-1} \partial_\sigma u_0 \rangle}{\langle \mathbf{u}_{\mathrm{ad}}, \mathbf{u}_1 \rangle} = \frac{-1}{\langle \mathbf{u}_{\mathrm{ad}}, \mathbf{u}_1 \rangle}$$

$$\beta_2 = \frac{-\langle D u_{\mathrm{ad}}, -\frac{1}{2} D^{-1} f''(u_0)[\partial_k u_0, \partial_k u_0] \rangle + \langle \mathbf{u}_{\mathrm{ad}}, \partial_\sigma \mathbf{u}_1 \rangle}{\langle \mathbf{u}_{\mathrm{ad}}, \mathbf{u}_1 \rangle}$$

$$= \frac{\langle u_{\mathrm{ad}}, 2k_0 D \partial_{k\sigma\sigma} u_0 - c_{\mathrm{g}}^0 \partial_{k\sigma} u_0 + D \partial_{\sigma\sigma} u_0 + \frac{1}{2} f''(u_0)[\partial_k u_0, \partial_k u_0] \rangle}{\langle \mathbf{u}_{\mathrm{ad}}, \mathbf{u}_1 \rangle}$$

$$= \frac{\frac{1}{2} \omega_{\mathrm{nl}}''(k_0)}{\langle \mathbf{u}_{\mathrm{ad}}, \mathbf{u}_1 \rangle}.$$

The denominator in both expressions is determined by the linear dispersion relation:

$$\langle \mathbf{u}_{\mathrm{ad}}, \mathbf{u}_1 \rangle = \langle k_0 D \partial_\sigma u_{\mathrm{ad}} + c_{\mathrm{g}}^0 u_{\mathrm{ad}}, -\partial_k u_0 \rangle + \langle D u_{\mathrm{ad}}, k_0 \partial_{k\sigma} u_0 + \partial_\sigma u_0 \rangle$$

$$= \langle u_{\mathrm{ad}}, 2k_0 D \partial_{k\sigma} u_0 + D \partial_\sigma u_0 \rangle = \frac{1}{2} \lambda_{\mathrm{lin}}''(0).$$

Alternatively, we can infer a relation between β_ω and β_2 using the following reverse argument. Upon inspection, we see that (8.13) has equilibria precisely when

$$\bar{\omega} = -\frac{\beta_2}{\beta_\omega} \kappa^2.$$

On the other hand, we know that the family of wave trains exists for frequencies $\omega = \omega_{\mathrm{nl}}(k)$. Since κ corresponds to the detuning of the wave number, we see that

$$\bar{\omega} = \frac{1}{2} \omega_{\mathrm{nl}}''(k_0) \kappa^2,$$

and comparing the equations for $\bar{\omega}$, we obtain the relation $\beta_2 = \frac{1}{2} \omega_{\mathrm{nl}}''(k_0) \beta_\omega$.

In summary, we have shown that the reduced vector field on the center manifold is of the form

$$\partial_\xi \theta = \kappa + \mathrm{O}(|\bar{\omega}| + |\kappa|^2)$$

(8.15) $\qquad \partial_\xi \kappa = \dfrac{1}{\frac{1}{2} \lambda_{\mathrm{lin}}''(0)} \left(\frac{1}{2} \omega_{\mathrm{nl}}''(k_0) \kappa^2 - \bar{\omega} \right) + \mathrm{O}(|\bar{\omega}|^2 + |\bar{\omega}\kappa| + |\kappa|^3)$

where neither of the remainder terms depends on θ. On the center manifold, we find heteroclinic solutions in the κ-equation which correspond precisely to the desired viscous shock waves. This finishes the proof of Theorem 4.10.

The statement of Remark 4.11 follows from the fact that a sign change of $\lambda_{\text{lin}}''(0)$ corresponds to replacing ξ by $-\xi$ in (8.15). Consequently, the stationary front given in (2.11) connects the asymptotic equilibria in the opposite order, and the relative group velocities, computed in the comoving frame, change their sign as well.

8.2. Proof of Theorem 4.12

To prove Theorem 4.12, we need to locate the spectrum of the linearization about a modulated wave.

We denote by $u_*(x - c_{\text{g}}^0 t, \omega_* t; \varepsilon)$ the solution that we constructed in Theorem 4.10, where we set $\varepsilon^2 = \omega_* - \omega_*^0$. Here, we assumed that $\omega_{\text{nl}}''(k_0) > 0$ (the case where the second derivative is negative can be handled in the same fashion). Furthermore, we denote by Ψ the time-$2\pi/\omega_*$-map of the reaction-diffusion system (4.1) and by $\Psi_*'(\varepsilon)$ the derivative of the period map with respect to the initial condition, evaluated at $u_*(x - c_{\text{g}}^0 t, \omega_* t; \varepsilon)$.

Our goal is to show that the spectrum of the linearized period map lies strictly inside the unit circle when we consider the operator on the exponentially weighted function spaces

$$L^2_{\eta_-,\eta_+}(\mathbb{R}) = \left\{ u \in L^2_{\text{loc}}(\mathbb{R});\ \|u\|_{L^2_{\eta_-,\eta_+}} < \infty \right\}$$

$$\|u\|^2_{L^2_{\eta_-,\eta_+}} = \int_{-\infty}^0 |u(x)\mathrm{e}^{\eta_- x}|^2\,\mathrm{d}x + \int_0^\infty |u(x)\mathrm{e}^{\eta_+ x}|^2\,\mathrm{d}x$$

defined in (4.47).

For each fixed choice of $\eta_\pm \in \mathbb{R}$, we consider $\Psi_*'(\varepsilon)$ as a bounded operator from $L^2_{\eta_-,\eta_+}$ into itself. For each $\Lambda \neq 0$ in the spectrum of $\Psi_*'(\varepsilon)$, we define its Floquet exponent λ by

$$\lambda = \frac{\omega_*}{2\pi} \log \Lambda.$$

We distinguish between values of λ in the Floquet point spectrum, where $\Psi_*'(\varepsilon) - \Lambda$ is Fredholm with index zero but not invertible, and values of λ in the essential Floquet spectrum, where $\Psi_*'(\varepsilon) - \rho$ is not Fredholm or Fredholm with nonzero index.

Before we state the first lemma, we recall from [49] that the function

$$(8.16)\quad \lambda_*(\nu;k) = (c_{\text{g}}^0 - c_{\text{p}}(k))\nu + \lambda_{\text{lin}}(\nu;k) = (c_{\text{g}}^0 - c_{\text{g}}(k))\nu + \frac{1}{2}\lambda_{\text{lin}}''(0;k)\nu^2 + \mathrm{O}(\nu^3),$$

which is defined and analytic in $\nu \in \mathbb{C}$, is the linear dispersion relation of the wave train with wave number k computed in the frame moving with speed c_{g}^0.

LEMMA 8.1. *For each choice of weights $\eta_\pm \in \mathbb{R}$, the essential Floquet spectrum of $\Psi_*'(\varepsilon)$ in $L^2_{\eta_-,\eta_+}$ is strictly to the left of the essential Floquet spectrum of the asymptotic wave trains computed in $L^2_{\eta_-,\eta_+}$, while the Floquet spectrum of the wave trains computed in $L^2_{\eta_-,\eta_+}$ is given by*

$$\lambda = \lambda_*(\mathrm{i}\ell - \eta_\pm; k_\pm), \qquad \ell \in \mathbb{R}.$$

8.2. PROOF OF THEOREM 4.12

In particular, any element λ in the essential Floquet spectrum of the viscous shocks satisfies

(8.17) $$\operatorname{Re}\lambda \leq \mp|\varepsilon|\sqrt{2\omega_{\mathrm{nl}}''(k_0)}\eta_\pm + (\lambda_{\mathrm{lin}}''(0) + \mathrm{O}(\varepsilon))\eta_\pm^2$$

provided $\eta_{\min} < \eta_- < 0 < \eta_+ < \eta_{\max}$ and

(8.18) $$\min\{|\eta_{\min}|, |\eta_{\max}|\} = |\varepsilon|\frac{\sqrt{2\omega_{\mathrm{nl}}''(k_0)}}{\lambda_{\mathrm{lin}}''(0)} + \mathrm{O}(\varepsilon^2),$$

where λ_{lin} denotes the linear dispersion relation of the wave train with wave number k_0.

For further reference, we remark that we obtain the optimal estimate

(8.19) $$\operatorname{Re}\lambda \leq -\varepsilon^2\frac{\omega_{\mathrm{nl}}''(k_0)}{2\lambda_{\mathrm{lin}}''(0)} + \mathrm{O}(\varepsilon^3)$$

for the essential spectrum when we substitute the optimal $\eta_\pm$ from (8.18) into (8.17).

PROOF. [49, Remark 2.9 and Proposition 2.10] assert that the linearization about the viscous shocks, computed in the frame moving with the speed c_g^0 of the shock, is Fredholm in the complement of the spectrum of the asymptotic wave trains. Also, the spectra of wave trains with wave numbers $k = k_\pm$, computed in the frame moving with speed c_g^0, are given by the dispersion relation (8.16)

$$\lambda_*(\nu; k_\pm) = (c_\mathrm{g}^0 - c_\mathrm{g}^\pm)\nu + \frac{1}{2}\lambda_{\mathrm{lin}}''(0; k_\pm)\nu^2 + \mathrm{O}(\nu^3)$$

where $c_\mathrm{g}^\pm = \omega_{\mathrm{nl}}'(k_\pm)$. Since

$$\varepsilon^2 = \omega_* - \omega_*^0 = \omega_{\mathrm{nl}}''(k_0)(k_\pm - k_0) + \mathrm{O}(|k_\pm - k_0|^3),$$

we see that

$$k_\pm - k_0 = \mp\frac{\sqrt{2}|\varepsilon|}{\sqrt{\omega_{\mathrm{nl}}''(k_0)}} + \mathrm{O}(\varepsilon^2)$$

where we recall that we assumed that $\omega_{\mathrm{nl}}''(k_0) > 0$. Therefore,

$$c_\mathrm{g}^0 - c_\mathrm{g}^\pm = -\omega_{\mathrm{nl}}''(k_0)(k_\pm - k_0) + \mathrm{O}(|k_\pm - k_0|^3) = \pm\sqrt{2\omega_{\mathrm{nl}}''(k_0)}|\varepsilon| + \mathrm{O}(\varepsilon^2).$$

We also have

$$\lambda_{\mathrm{lin}}''(0; k_\pm) = \lambda_{\mathrm{lin}}''(0) + \mathrm{O}(\varepsilon)$$

with $\lambda_{\mathrm{lin}}''(0; k_\pm) > 0$ by Hypothesis 4.2. Thus, we see that

$$\lambda(\nu; k_\pm) = \left[\pm\sqrt{2\omega_{\mathrm{nl}}''(k_0)}|\varepsilon| + \mathrm{O}(\varepsilon^2)\right]\nu + [\lambda_{\mathrm{lin}}''(0) + \mathrm{O}(\varepsilon)]\nu^2 + \mathrm{O}(\nu^3).$$

Substituting $\nu = -\eta_\pm + i\ell$, we obtain

$$\begin{aligned}\operatorname{Re}\lambda(\nu; k_\pm) &\leq \operatorname{Re}\lambda(-\eta_\pm; k_\pm) \\ &= \left[\mp\sqrt{2\omega''(k_0)}|\varepsilon| + \mathrm{O}(\varepsilon^2)\right]\eta_\pm + [\lambda_{\mathrm{lin}}''(0) + \mathrm{O}(\varepsilon)]\eta_\pm^2 + \mathrm{O}(\eta_\pm^3)\end{aligned}$$

which is strictly negative provided $\eta_{\min} < \eta_- < 0 < \eta_+ < \eta_{\max}$ and

(8.20) $$\min\{|\eta_{\min}|, |\eta_{\max}|\} \leq |\varepsilon|\frac{\sqrt{2\omega_{\mathrm{nl}}''(k_0)}}{\lambda_{\mathrm{lin}}''(0)} + \mathrm{O}(\varepsilon^2)$$

for ε sufficiently close to zero. $\square$

LEMMA 8.2. *For each $\varepsilon > 0$ sufficiently small, the Floquet point spectrum of the viscous shocks is contained strictly in the open left half-plane.*

Before we give the proof of the lemma, we show that the two lemmata imply Theorem 4.12.

PROOF OF THEOREM 4.12. Define $\mathcal{Y} = H^1_{\eta_-,\eta_+}$ where we choose the exponential weights as described in Theorem 4.12. Lemma 8.1 and 8.2 assert that the Floquet spectrum of the linearized period map $\Psi'_*(\varepsilon)$ associated with the viscous shocks on $\mathcal{Y}$ is contained in the open left half-plane. Therefore, $[\Psi'_*(\varepsilon)]^N$ is a contraction for some sufficiently large integer $N \gg 1$. Next, note that the nonlinearity is actually smooth when considered as map from $\mathcal{Y}$ into itself, since the exponential weights enforce functions to be localized. Thus, the variation-of-constants formula shows that the nonlinear period map is close to the linearized period map in the $\mathcal{C}^1$-topology if we restrict them to a sufficiently small neighbourhood of the viscous shock. As a consequence, the iterated nonlinear period map is a contraction in a sufficiently small neighbourhood of the viscous shock which proves its nonlinear stability in $\mathcal{Y}$. This proves Theorem 4.12. □

PROOF OF LEMMA 8.2. We have to prove that the operator $\Psi'_* - \lambda$ cannot have exponentially localized functions in its null space for any λ near zero. To show this, we write the Floquet eigenvalue problem as a differential equation

$$\begin{aligned}(8.21) \qquad \partial_\xi u &= v \\ \partial_\xi v &= -D^{-1}[-\omega_* \partial_\tau u - \lambda u + c_g^0 v + f'(u_*(\xi,\tau;\varepsilon))u]\end{aligned}$$

in the spatial variable ξ. Theorem 4.10 shows that

$$u_*(\xi,\tau;\varepsilon) = u_0(\tau - \xi - \theta_*(\xi)), \qquad \partial_\xi \theta_*(\xi) = \kappa_*(\xi), \qquad \kappa_*(\xi) = \varepsilon\kappa_0 \tanh(\varepsilon\xi)$$

where the constant κ_0 can be computed easily in terms of linear and nonlinear dispersion relation.

Our strategy is similar to the one we used to prove existence of viscous shocks. First, for $\lambda \approx 0$, we can reduce (8.21) to a two-dimensional nonautonomous center manifold that contains all bounded solutions to (8.21). To compute the vector field on the center manifold, we choose appropriate coordinates on it. We therefore pass to the comoving frame $\sigma = \tau - k_0 \xi$ and parametrize solutions by

$$(8.22) \qquad \mathbf{u} = \tilde{\theta}\partial_\sigma \mathbf{u}_0 + \tilde{\kappa}\mathbf{u}_1 - \tilde{\theta}\kappa_* \mathbf{u}_1.$$

For $\lambda = 0$, the coordinates (8.22) correspond to the linearization of the coordinates used to construct the viscous shock. In particular, using these coordinates, we recover the linearization of (8.15) at the viscous shock. Thus, we need to calculate only the reduced term that corresponds to λu_σ. This expression, however, enters the reduced equation in the same form as the term $\mathcal{N}$ enters the nonlinear problem, at least to leading order. Therefore, we end up with a reduced eigenvalue problem

$$\begin{aligned}(8.23) \qquad \partial_\xi \tilde{\theta} &= \tilde{\kappa} + \mathrm{O}(\varepsilon^2 + |\varepsilon\lambda|) \\ \partial_\xi \tilde{\kappa} &= \frac{1}{\frac{1}{2}\lambda''_{\mathrm{lin}}(0)}\left[\lambda\tilde{\theta} + \omega''_{\mathrm{nl}}(k_0)\kappa_*(\xi)\tilde{\kappa}\right] + \mathrm{O}(|\varepsilon\lambda| + |\varepsilon^2\tilde{\kappa}|).\end{aligned}$$

It is not hard to see that bounded solutions occur for $\mathrm{Re}\,\lambda \geq 0$ only in the scaling $\lambda = \varepsilon^2 \tilde{\lambda}$ and $X = \varepsilon\xi$, since we recover the heat equation outside this scaling which does not have bounded unstable eigenfunctions outside a bounded disk. In the scaled coordinates, the eigenvalue problem becomes

$$\text{(8.24)} \qquad \frac{1}{2}\lambda_{\text{lin}}''(0)\partial_{XX}\tilde{\theta} - \omega_{\text{nl}}''(k_0)\tanh(X)\partial_X\tilde{\theta} = \tilde{\lambda}\tilde{\theta} + \mathrm{O}(\varepsilon).$$

This eigenvalue problem arises also through the linearization about shocks in the eikonal equation

$$\partial_T\tilde{\theta} = \partial_{XX}\tilde{\theta} + (\partial_X\tilde{\theta})^2$$

which can be viewed as the integrated form $q = \partial_X\tilde{\theta}$ of the Burgers equation

$$\partial_T q = \partial_{XX} q + \partial_X(q^2).$$

Since the Evans function for the linearization about viscous shocks in the Burgers equation does not have zeros in a bounded neighbourhood of the origin except at $\tilde{\lambda} = 0$, the only exponentially localized eigenfunction is given by the derivative of the shock profile at $\tilde{\lambda} = 0$. This solution, however, is not exponentially localized as a solution to (8.24), since $\partial_X\tilde{\theta}(X)$ converges to nonzero constants as $X \to \pm\infty$. Therefore, the Evans function for the eigenvalue problem (8.24) with $\varepsilon = 0$ does not vanish in $\operatorname{Re}\tilde{\lambda} \geq -a$ for some $a > 0$. A continuity argument with respect to ε shows the absence of point spectrum and concludes the proof of Lemma 8.2. □

CHAPTER 9

Existence of shocks in the long-wavelength limit

9.1. A lattice model for weakly interacting pulses

In this section, we investigate the long-wavelength limit of wave trains. We assume that the wave-train profile converges to a localized pulse as the wave number k tends to zero. Thus, the wave train itself resembles an infinite chain of localized pulses that interact weakly with each other. We are then interested in finding the analogues of viscous shocks for the resulting lattice equation that describes weakly interacting pulses. To be specific, we consider again the reaction-diffusion system (4.1):

(9.1) $$\partial_t u = D\partial_{xx} u + f(u).$$

We begin by motivating the lattice equation that we are going to investigate. Thus, assume that

$$u(x,t) = h(x - c_\mathrm{p} t)$$

is a localized travelling-wave solution to (9.1) with phase speed c_p so that $h(\zeta) \to 0$ as $\zeta \to \pm\infty$. In particular, $h(\zeta)$ is a homoclinic orbit of the travelling-wave ODE

(9.2) $$\frac{\mathrm{d}}{\mathrm{d}\zeta} \begin{pmatrix} u \\ v \end{pmatrix} = \begin{pmatrix} v \\ D^{-1}[c_\mathrm{p} v - f(u)] \end{pmatrix}.$$

We shall then be interested in solutions to (9.1) of the approximate form

(9.3) $$u(x,t) \approx \sum_{j=-\infty}^{\infty} h(x - c_\mathrm{p} t + \zeta_j(t))$$

where the time-dependent positions $\zeta_j(t)$ account for the interaction of individual pulses. We assume that the pulses are widely spaced to that, at least initially, $\zeta_{j+1} - \zeta_j \gg 1$ for all $j \in \mathbb{Z}$.

Next, we want to write down an ODE that gives the evolution of the positions $\zeta_j(t)$ of the pulses in the train (9.3). We may expect that the pulses interact, at least to leading order, only with their nearest neighbours, so that the equation for ζ_j depends only on the distances $\zeta_{j+1} - \zeta_j$ and $\zeta_j - \zeta_{j-1}$ of the jth pulse to its nearest neighbours. If the localized pulse $h(\zeta)$ decays to zero exponentially, then the function given in (9.3) is a solution to the reaction-diffusion equation up to an exponentially small error that arises due to the overlapping tails. Thus, we expect that the equation of motion for the pulse positions should be exponentially small in the distances between consecutive pulses. For a finite number of pulses, equations of this kind have indeed been formally derived, for instance, in [15]. Rigorous results that validate these equations for a finite number of pulse can be found in [14, 47], while their validity for an infinite number of pulses has recently been established in [58].

To write down the ODE, we assume that the linearization of (9.2) about the equilibrium $(u,v) = 0$ is hyperbolic and that there is a unique simple eigenvalue closest to the imaginary axis. We may then assume that this eigenvalue is stable (the other case can be treated in exactly the same fashion), and we denote it by $\nu = -b < 0$ with $b > 0$. The resulting lattice equation is

$$\frac{d\zeta_j}{dt} = a e^{-b(\zeta_j - \zeta_{j-1})}, \qquad j \in \mathbb{Z} \tag{9.4}$$

where we assume that $a \neq 0$. The equation for the jth pulse depends only on the distance to the pulse behind, but not the pulse ahead. This reflects the fact that, while the jth pulse sees the tails of both neighbouring pulses, the tail of the pulse behind decays much slower due to our assumption that the stable eigenvalue is closest to the imaginary axis. Note also that we omit all remainder terms in (9.4). We summarize our assumptions on the coefficients that appear in (9.4) in the following hypothesis.

HYPOTHESIS 9.1. *We assume that $a \neq 0$ and $b > 0$ in (9.4).*

We remark that wave trains with large spatial period L, or small wave number $k = 1/L$, correspond to the solutions $\zeta_j(t) - \zeta_{j-1}(t) = L$ of (9.4) for $j \in \mathbb{Z}$ where $L \gg 1$ is fixed. These wave trains have phase speed $c_p(L) = a \exp(-bL)$.

Since we are interested in finding fronts that connect different wave trains as $j \to \pm\infty$, it is more convenient to use the distances

$$L_j(t) = \zeta_j(t) - \zeta_{j-1}(t) \tag{9.5}$$

instead of the positions ζ_j as variables. We then obtain the equivalent lattice equation

$$\frac{dL_j}{dt} = a \left(e^{-bL_j} - e^{-bL_{j-1}} \right), \qquad j \in \mathbb{Z} \tag{9.6}$$

for the distances $L_j(t)$ between the jth and the $(j-1)$th pulse. Viscous shocks with nonzero speed c_* that connect different asymptotic wave trains with periods $L_\pm$ correspond then to travelling-wave solutions to (9.6) of the form

$$L_j(t) = L_*(j - c_* t), \qquad j \in \mathbb{Z} \tag{9.7}$$

with $L_*(\xi) \to L_\pm$ as $\xi \to \pm\infty$, where $L_*(\xi)$ is a given function, defined for $\xi \in \mathbb{R}$, that describes the profile of the shock. Substituting the above ansatz into (9.6), we obtain the delay equation

$$\frac{dL}{d\xi}(\xi) = -\frac{a}{c_*} \left(e^{-bL(\xi)} - e^{-bL(\xi - 1)} \right), \qquad \xi \in \mathbb{R} \tag{9.8}$$

for the profile $L_*(\xi)$ where $\xi = j - c_* t$. We can now state the main result of this section.

THEOREM 9.2. *Assume that Hypothesis 9.1 is met. For any values $L_+ > L_- > 0$, there exist a constant c_* and a strictly monotonically increasing solution $L_*(\xi)$ of (9.8) such that $L_*(\xi) \to L_\pm$ as $\xi \to \pm\infty$ and $\mathrm{sign}(ac_*) > 0$. If $L_- > L_+ > 0$, then a solution $L_*(\xi)$ with the above properties does not exist for any value of c_*.*

Theorem 9.2 is somewhat stronger than the corresponding Theorem 4.10 for reaction-diffusion systems, since it is not required in Theorem 9.2 that $|L_+ - L_-|$ is small.

9.2. Proof of Theorem 9.2

It will be convenient to use the new variable r defined by

$$(9.9) \qquad r = e^{-bL}, \qquad L = -\frac{\ln r}{b}$$

instead of L so that $L > 0$ corresponds to $0 < r < 1$ (recall $b > 0$). Equation (9.8) then becomes

$$\frac{dr}{d\xi}(\xi) = \frac{ab}{c_*} r(\xi)(r(\xi) - r(\xi - 1)), \qquad \xi \in \mathbb{R},$$

which we write as

$$(9.10) \qquad \frac{dr}{d\xi}(\xi) = Ar(\xi)(r(\xi) - r(\xi - 1)), \qquad \xi \in \mathbb{R},$$

for $r > 0$ where

$$A = \frac{ab}{c_*}$$

is arbitrary. Any constant function $r(\xi) = r_0$ satisfies (9.10). The characteristic equation of the linearization

$$(9.11) \qquad \frac{ds}{d\xi}(\xi) = Ar_0(s(\xi) - s(\xi - 1))$$

about $r(\xi) = r_0$ is obtained by seeking solutions to (9.11) of the form $s(\xi) = \exp(\lambda \xi)$ and is therefore given by

$$(9.12) \qquad \Delta(\lambda) = \lambda - Ar_0\left(1 - e^{-\lambda}\right) = 0.$$

We then have the following result.

LEMMA 9.3 ([12]). *Assume that Hypothesis 9.1 is met and fix $r_0 > 0$. The characteristic equation (9.12) has the root $\lambda = 0$. In addition, there is precisely one other real root: this root is positive for $Ar_0 > 1$, negative for $Ar_0 < 1$, and zero for $Ar_0 = 1$. All remaining solutions of (9.12) have nonzero imaginary part and strictly negative real part regardless of the value of A.*

PROOF. The assertions are a consequence of

$$\Delta(0) = 0, \qquad \frac{d\Delta}{d\lambda}(0) = 1 - Ar_0, \qquad \lim_{\lambda \to \infty} \operatorname{sign} \Delta(\lambda) = 1$$

taken together with [12, Theorems 3.1, 3.2 and 3.12]. $\square$

The root $\lambda = 0$ corresponds, of course, to the line of equilibria of (9.10) given by $r(\xi) = r_0$ where $r_0 > 0$ is arbitrary. Interpreting (9.10) as a dynamical system on the function space $C^0([-1, 0])$, see [12], and using the invariant-manifold results stated and proved in [12, Chapters VIII and IX], we can therefore conclude the following from Lemma 9.3: The unstable manifold of the equilibrium $r(\xi) = r_0$ is one-dimensional if $Ar_0 > 1$ and has dimension zero for $Ar_0 < 1$. Analogously, the stable manifold of $r(\xi) = r_0$ has codimension two if $Ar_0 > 1$ and codimension one if $Ar_0 < 1$. The line $r = r_0 \in \mathbb{R}^+$ of equilibria forms the center manifold which has dimension one at points where $Ar_0 \neq 1$.

In particular, since $0 < r_0 < 1$ for any equilibrium consistent with (9.9), we see that $A > 0$ is a necessary condition to obtain a heteroclinic orbit that connects an equilibrium r_- to r_+. Thus, we will only consider $A > 0$ from now on and define

$$(9.13) \qquad \ell = \frac{r}{A} > 0$$

so that (9.10) becomes

$$\frac{d\ell}{d\xi}(\xi) = \ell(\xi)(\ell(\xi) - \ell(\xi - 1)). \tag{9.14}$$

LEMMA 9.4. *If $\ell(\xi)$ satisfies (9.14) for $\xi \geq 0$, and $\ell(\xi) > 0$ is strictly monotone on $[-1, 0]$, then $\ell(\xi)$ is strictly monotone for $\xi \geq -1$ as long as $\ell(\xi) > 0$.*

PROOF. If not, take the smallest value of $\xi \geq 0$ for which $\ell'(\xi) = 0$, while $\ell(\xi) > 0$. It is then straightforward to obtain a contradiction to (9.14). □

The next lemma states that every $\ell_- > 1$ connects to some $\ell_+ > 0$ via a heteroclinic solution to (9.14).

LEMMA 9.5. *For each $\ell_- > 1$, there exists a solution $\ell_*(\xi)$ to (9.14) in the one-dimensional unstable manifold of ℓ_- such that $\ell_*(\xi)$ decreases monotonically for $\xi \in \mathbb{R}$ and $\ell_*(\xi) \to \ell_+$ for some $\ell_+ > 0$.*

PROOF. The tangent vector to the unstable manifold of ℓ_- is equal to $\exp(\lambda \xi)$ for some $\lambda > 0$. In particular, the solution $\ell_*(\xi)$ in the unstable manifold that corresponds to $\ell_- - \varepsilon \exp(\lambda \xi) + O(\varepsilon^2)$ for sufficiently small $\varepsilon > 0$ is strictly monotonically decreasing in ξ for all $\xi \ll -1$. On account of Lemma 9.4, it therefore suffices to show that $\ell_*(\xi)$ is bounded away from zero in order to prove the lemma. To this end, define $p(\xi) = 1/\ell(\xi)$ so that $p(\xi) = 1/\ell_*(\xi)$ satisfies the equation

$$p'(\xi) = \frac{p(\xi) - p(\xi - 1)}{p(\xi - 1)}.$$

We shall show that the monotonically increasing $p(\xi)$ is bounded as a function of ξ. If not, then we certainly have that $p(\xi) > 4$ for all $\xi \geq -1$, say. Integrating the above delay equation, we therefore obtain

$$\begin{aligned}
\Delta(\xi) \; := \; p(\xi) - p(\xi - 1) &= \int_{\xi-1}^{\xi} \frac{p(\zeta) - p(\zeta - 1)}{p(\zeta - 1)} \, d\zeta \\
&\leq \frac{1}{4} \sup_{\zeta \in [\xi-1, \xi]} [p(\zeta) - p(\zeta - 1)] = \frac{1}{4} \sup_{\zeta \in [\xi-1, \xi]} \Delta(\zeta)
\end{aligned}$$

for $\xi \geq 0$. Taking the supremum on both sides gives

$$\begin{aligned}
\sup_{\xi \in [\tau, \tau+1]} \Delta(\xi) &\leq \frac{1}{4} \sup_{\xi \in [\tau, \tau+1]} \left(\sup_{\zeta \in [\xi-1, \xi]} \Delta(\zeta) \right) \leq \frac{1}{4} \sup_{\xi \in [\tau-1, \tau+1]} \Delta(\xi) \\
&\leq \frac{1}{4} \left(\sup_{\xi \in [\tau-1, \tau]} \Delta(\xi) + \sup_{\xi \in [\tau, \tau+1]} \Delta(\xi) \right)
\end{aligned}$$

so that

$$M_j := \sup_{\xi \in [j, j+1]} \Delta(\xi) < \frac{1}{2} \sup_{\xi \in [j-1, j]} \Delta(\xi) = \frac{1}{2} M_{j-1}$$

for all $j \geq 0$. As a result, we get

$$p(j) = p(1) + \sum_{i=0}^{j} \Delta(i) \leq p(1) + \sum_{i=0}^{j} M_i \leq p(1) + \sum_{i=0}^{j} \frac{M_1}{2^i} \leq p(1) + 2M_1$$

which proves that $p(\xi)$ is indeed bounded independently of ξ. □

For each $\ell_- > 1$, we denote by $\ell_+ = H(\ell_-)$ the constant solution for which $\lim_{\xi \to \infty} \ell_*(\xi) = \ell_\pm$, where $\ell_*(\xi)$ is the heteroclinic orbit obtained in Lemma 9.5.

LEMMA 9.6. *The function $H(\ell_-)$ is continuous in $\ell_- > 1$, has values in $(0, 1]$, and satisfies*

(9.15) $$\frac{H(\ell_-)}{\ell_-} \longrightarrow \begin{cases} 0 & \ell_- \to \infty \\ 1 & \ell_- \to 1 \end{cases}$$

Before we prove this lemma, we show how it implies Theorem 9.2. Choose $L_+ > L_- > 0$, then we need to find numbers $\ell_- > 1$ and A such that

$$H(\ell_-) = \frac{1}{A} e^{-bL_+}, \qquad \ell_- = \frac{1}{A} e^{-bL_-}.$$

Thus, $A = e^{-bL_-}/\ell_-$, and it remains to find $\ell_- > 1$ such that

$$\frac{H(\ell_-)}{\ell_-} = e^{-b(L_+ - L_-)}.$$

Lemma 9.6 together with $L_+ - L_- > 0$ implies that such an ℓ_- exists. Thus, it suffices to prove Lemma 9.6.

PROOF OF LEMMA 9.6. We first prove that $H(\ell_-) \leq 1$ for all $\ell_- > 1$ (note that H is well defined and positive by Lemma 9.5). Observe that the derivative $\ell'_*(\xi)$ satisfies the variational equation

(9.16) $$\frac{d\ell}{d\xi}(\xi) = [2\ell_*(\xi) - \ell_*(\xi - 1)]\ell(\xi) - \ell_*(\xi)\ell(\xi - 1).$$

about the connecting orbit $\ell_*(\xi)$. The results in [9] imply that (9.16) does not admit any nontrivial small solutions, that is, there are no nonzero solutions to (9.16) that decay faster than any given exponential. If $H(\ell_-) > 1$, then this fact implies that

$$\ell'_*(\xi) = a^{-\lambda\xi} + O\left(^{-2\lambda\xi}\right)$$

for some $a \neq 0$ and some root λ of $\Delta(\lambda)$, defined in (9.12), with $\text{Re}\,\lambda < 0$. By Lemma 9.3, any such λ has nonzero imaginary part though, contradicting monotonicity of $\ell_*(\xi)$ as stated in Lemma 9.4. Therefore, we have $H(\ell_-) \leq 1$ for all $\ell_- > 1$.

Using this restriction on the range of H, continuity of H follows from the unstable-manifold theorem for delay equations [12]. As a consequence of the above facts, we see that $H(\ell_-)/\ell_- \to 0$ as $\ell_- \to \infty$.

It therefore remains to prove that $H(\ell_-)/\ell_- \to 1$ as $\ell_- \to 1$. Hence, we consider the two-dimensional center manifold of (9.14),

$$\frac{d\ell}{d\xi}(\xi) = \ell(\xi)(\ell(\xi) - \ell(\xi - 1)),$$

near $\ell = 1$. Using the results in [12, Chapter IX.10], we see that the vector field on the center manifold is given by

(9.17) $$\begin{aligned} x' &= y\left(1 + \frac{2x}{3} + O(x^2 + y^2)\right) \\ y' &= y\left(2x + O(x^2 + y^2)\right) \end{aligned}$$

where the coordinates x and y correspond to the eigenfunction $\ell(\xi) = 1$ and the generalized eigenfunction $\ell(\xi) = \xi$. We also used the fact that the line $y = 0$ consists of equilibria. Introducing the new variable

$$z = x' = y\left(1 + \frac{2x}{3} + O(x^2 + y^2)\right),$$

(9.17) becomes

(9.18) $$\begin{aligned} x' &= z \\ z' &= 2z\left(x + \frac{z}{3} + O(x^2 + z^2)\right). \end{aligned}$$

The equilibria with $x > 0$ have a one-dimensional unstable manifold. The solutions inside these manifolds for which $z < 0$ will cross the z-axis at a finite distance. We shall construct a trapping region that shows that each such solution converges to an equilibrium with $x < 0$. Indeed, consider the line

$$z = T(x) = -\varepsilon^2\left(1 + \frac{x}{2\varepsilon}\right) =: -\varepsilon^2 w$$

where $-2\varepsilon \leq x \leq 0$ and therefore $w \in [0, 1]$. We compute

$$\begin{pmatrix} -\frac{dT}{dx} \\ 1 \end{pmatrix} \cdot \begin{pmatrix} x' \\ z' \end{pmatrix} = \frac{\varepsilon^3 y}{2}\left[5 - 4y + O(\varepsilon)\right] > 0$$

which shows that solutions in the unstable manifold of equilibria with $x > 0$ close to zero converge to equilibria with $x < 0$ that are also close to zero. Interpreting these results for the original equation proves that $H(\ell_-)/\ell_- \to 1$ as $\ell_- \to 1$. □

CHAPTER 10

Applications

10.1. The FitzHugh–Nagumo equation

The FitzHugh–Nagumo equation is given by

$$\begin{aligned}
(10.1) \qquad \partial_t u &= \partial_{xx} u + u(1-u)(u-a) - w \\
\partial_t w &= \varepsilon(u - \gamma w),
\end{aligned}$$

for $x \in \mathbb{R}$, where $\gamma \geq 0$ and $a \in (0, \frac{1}{2})$ are fixed. This equation is a simplification of the Hodgkin–Huxley equation that models the propagation of impulses in nerve axons. We are interested in travelling waves $(u, w)(x, t) = (u, w)(x - ct)$.

It has been shown in [21] that (10.1) exhibits a localized pulse with positive speed for all sufficiently small $0 < \varepsilon \ll 1$. As shown in [24, 57], this pulse, which we refer to as the fast pulse, is nonlinearly stable. Each fast pulse is accompanied by a family of wave trains with arbitrarily large period.

THEOREM 10.1 ([48, Theorem 21]). *For each fixed a in the interval $(0, \frac{1}{2})$, there exists a number $\varepsilon_* = \varepsilon_*(a)$ with the following property. For every ε with $0 < \varepsilon < \varepsilon_*$, there is an $L_* = L_*(\varepsilon)$ so that the fast pulse to the FitzHugh–Nagumo system is accompanied by periodic wave trains with period L for any $L > L_*$, and all these wave trains are spectrally stable.*

In fact, using the quantities ν^s, V^s, W^s and M from [48, §6.2], we may define the constants b and Γ via

$$b = -\nu^s, \qquad \Gamma = \frac{\langle V^s, W^s \rangle}{M}$$

so that

$$(10.2) \qquad \lambda(\nu) = b\Gamma \left(e^{\nu L} - 1 \right) e^{-bL}, \qquad c(L) = c_\infty - \Gamma e^{-bL}$$

for all $L \gg 1$, where $c_\infty > 0$ and $c(L)$ denote the phase velocities of the fast pulse and the wave trains with period L, respectively. From Remark 4.5, we obtain that

$$c_g = c(L) - Lc'(L) = c(L) - bL\Gamma e^{-bL} < c(L) < c_\infty, \qquad \operatorname{sign} \omega_{\mathrm{nl}}''(k) = \operatorname{sign} c''(L) < 0.$$

Using geometric singular perturbation theory, the results mentioned above carry over to the modified FitzHugh–Nagumo equation

$$\begin{aligned}
(10.3) \qquad \partial_t u &= \partial_{xx} u + u(1-u)(u-a) - w \\
\partial_t w &= \delta^2 \partial_{xx} v + \varepsilon(u - \gamma w),
\end{aligned}$$

with small diffusion added to the second equation, provided the diffusion coefficient $\delta > 0$ is chosen sufficiently small so that $0 < \delta \ll \varepsilon \ll 1$. Hence, the theory developed in the preceding sections can be applied to the FitzHugh-Nagumo system (10.3). In particular, weak viscous-shock interfaces of (10.3) travel to the right at a smaller speed than the wave trains, and they connect wave trains with larger period at $x = -\infty$ to wave trains with smaller period at $x = \infty$.

Lastly, we remark that Eszter [16] investigated spectral and nonlinear stability of periodic wave trains to the FitzHugh–Nagumo system (10.1) in a different regime: he first fixed the period L of a singular spatially-periodic wave train and then varied $\varepsilon > 0$ near zero with $\varepsilon < \varepsilon_*(L)$; the maximal allowed value $\varepsilon_*(L)$ will tend to zero as the period L tends to infinity.

10.2. The weakly unstable Taylor–Couette problem

In §3, we asserted that solutions to the Burgers equation can indeed, as expected, be used to approximate the dynamics of modulated wave trains in the complex cubic Ginzburg–Landau equation. On the other hand, it is well known that the Ginzburg–Landau equation itself approximates the dynamics near onset of far more complex pattern-forming systems. It is the purpose of this section to illustrate this connection by investigating the Taylor–Couette problem close to the first instability of the stationary Couette flow.

We strongly expect that the results we established for reaction–diffusion systems are also true for hydrodynamical stability problems such as the Taylor–Couette problem but have not yet embarked on the proofs.

The Taylor–Couette problem [7] consists of finding the velocity field of a viscous incompressible fluid between two rotating concentric cylinders. This system has a stationary solution, the so-called Couette flow, that has purely azimuthal form, so that the streamlines are concentric circles. It is known that the Couette flow is asymptotically stable for sufficiently small Reynolds number $\mathcal{R}$ and destabilizes for larger Reynolds numbers. Mathematically, the fluid flow can be described by the incompressible Navier–Stokes equation with no-slip boundary conditions.

To set up the problem, we denote by R_i and R_o the inner and outer radii of the two concentric cylinders, with the obvious assumption $R_i < R_o$, and by Ω_i and Ω_o their angular velocities. We write ν for the viscosity coefficient of the fluid. It is then convenient to introduce the nondimensional parameters

$$\Omega := \Omega_o/\Omega_i, \qquad \eta := R_i/R_o, \qquad \mathcal{R} := R_i \Omega_i (R_o - R_i)/\nu$$

that fully describe the system, where $\mathcal{R}$ is called the Reynolds number. The annular planar cross-section between the cylinders is denoted by Σ, so that the fluid fills the three-dimensional volume $Q = \mathbb{R} \times \Sigma$. Thus, in cylindrical coordinates (x, r, φ), the domain Q is defined by $x \in \mathbb{R}$, $\eta/(1-\eta) < r < 1/(1-\eta)$, and $\varphi \in \mathbb{R}/2\pi\mathbb{Z}$. The Cartesian coordinates in the annular cross-section Σ are denoted by $z = (z_1, z_2) \in \Sigma \subset \mathbb{R}^2$.

The stationary Couette fluid flow is given by

$$U_{\mathrm{Cou}}(x, r, \varphi) = \begin{pmatrix} U_{(x)} \\ U_{(r)} \\ U_{(\varphi)} \end{pmatrix} = \left(Ar + \frac{B}{r}\right) \begin{pmatrix} 0 \\ 0 \\ 1 \end{pmatrix}$$

$$A = \frac{\Omega - \eta^2}{\eta(1+\eta)}, \qquad B = \frac{(1-\Omega)\eta}{(1-\eta)(1-\eta^2)},$$

where $(U_{(x)}, U_{(r)}, U_{(\varphi)})$ denote the cylindrical coordinates of the vector U. The above fluid flow satisfies the Navier–Stokes equation on Q with no-slip boundary conditions on ∂Q and is, in fact, exponentially stable for sufficiently small Reynolds numbers $\mathcal{R}$. The deviation (U, P) from the Couette flow U_{Cou} satisfies the Navier–Stokes equation

(10.4) $\quad \partial_t U = \Delta U - \mathcal{R}[(U_{\text{Cou}} \cdot \nabla)U + (U \cdot \nabla)U_{\text{Cou}} + (U \cdot \nabla)U] - \nabla P$
$\quad\quad\quad \nabla \cdot U = 0$

with no-slip boundary conditions $U = 0$ at $r = \eta/(1 - \eta)$ and at $r = 1/(1 - \eta)$. To solve this equation uniquely for the velocity U and pressure gradient ∇P, we need to add the flux condition

$$[U_{(x)}]_\Sigma = \frac{1}{|\Sigma|} \int_{z \in \Sigma} U_{(x)}(x, z) \, \mathrm{d}z = 0.$$

We refer to [7] for more details.

In the (U, P) variables, the Couette flow corresponds to $(U, P) \equiv 0$ which is a solution for all $\mathcal{R}$. This trivial branch of solutions becomes unstable when the Reynolds number $\mathcal{R}$ exceeds a certain threshold value which we denote by $\mathcal{R}_c$. The translation invariance of (10.4) in the x-direction implies that the linearization of (10.4) about $(U, P) = 0$ has continuous spectrum given by dispersion curves $\lambda = \lambda_n(\ell)$ for $n \in \mathbb{N}$ with associated eigenmodes of the form

$$e^{\lambda_n(\ell)t} e^{\mathrm{i}\ell x} U_n(\ell, z), \qquad z \in \Sigma$$

where $\ell \in \mathbb{R}$ and $U_n(\ell, z) \in \mathbb{C}^3$. We may order the dispersion curves so that $\operatorname{Re} \lambda_n \geq \operatorname{Re} \lambda_{n+1}$ for all $n \in \mathbb{N}$. The instability scenario is then as follows.

For η close to one, there exists a number Ω_b with the following property. If we fix $\Omega > \Omega_b$, then the real-valued curve $\ell \mapsto \lambda_1(\ell)$ crosses the imaginary axis from left to right as $\mathcal{R}$ increases though $\mathcal{R}_c$. On the other hand, if we fix $\Omega < \Omega_b$, then the two complex-conjugated curves $\ell \mapsto \lambda_1(\ell)$ and $\ell \mapsto \lambda_2(\ell) = \overline{\lambda_1(\ell)}$ cross the imaginary axis at some nonzero wave number $\ell = \ell_c \neq 0$ as $\mathcal{R}$ passes though $\mathcal{R}_c$. In both cases, each remaining dispersion curve is strictly bounded away from the imaginary axis. We refer to the first case as PRI and to the second case as PRII. In the following, we focus on PRII.

To analyse the resulting bifurcation for PRII, we introduce the small parameter $\varepsilon^2 = \mathcal{R} - \mathcal{R}_c$ and make the ansatz

(10.5) $\quad U_{\text{approx}} = \varepsilon A(\varepsilon(x - c_g t), \varepsilon^2 t) e^{\mathrm{i}\ell_c x + \mathrm{i}\omega_c t} U_1(\ell_c, z) + \text{c.c.}$

where

$$c_g = \frac{\mathrm{d} \operatorname{Im} \lambda_1}{\mathrm{d} \ell}(\ell_c), \qquad \omega_c = \operatorname{Im} \lambda_1(\ell_c).$$

Using this ansatz, the Ginzburg–Landau equation

(10.6) $\quad \partial_T A = c_1 \partial_{XX} A + c_2 A - c_3 |A|^2 A$

can be derived for the complex-valued amplitude $A = A(X, T)$ for certain complex coefficients $c_j \in \mathbb{C}$. It has been proved in [55] that the approximation of the Taylor–Couette system by the above Ginzburg–Landau equation is valid over the natural time scale:

THEOREM 10.2 ([55]). *For each choice of positive numbers C_1 and T_0, there exist constants C_2 and $\varepsilon_0 > 0$ such that the following is true for all $0 < \varepsilon < \varepsilon_0$. If $A \in \mathcal{C}^0([0, T_0], H^3_{\mathrm{ul}})$ is a solution to the Ginzburg–Landau equation (10.6) such that*

$$\sup_{T \in [0, T_0]} \|A(T)\|_{H^3_{\mathrm{ul}}} < C_1,$$

then there exists a solution U of the Taylor–Couette problem (10.4) with
$$\sup_{t\in[0,T_0/\varepsilon^2]} \|U(t) - U_{\text{approx}}(t)\|_{H^2_{\text{ul}}} \leq C_2\varepsilon^2$$
where U_{approx} has been defined in (10.5).

REMARK 10.3. In the parameter region PRII of interest to us, a system of coupled Ginzburg–Landau equations can be derived for the amplitudes A_1 and A_2 corresponding to the curves λ_1 and λ_2 of eigenvalues. Since these equations decouple when one of the amplitudes is set to be zero, we again obtain a family of solutions that can be described by a single Ginzburg–Landau equation (see [55]).

The Ginzburg–Landau equation (10.6) can be put into the normal form (3.11). Doing this in the region PRI, we obtain the coefficients $\alpha = \beta = 0$. In the case PRII, however, we obtain nonzero coefficients $\alpha, \beta \neq 0$ and therefore a nontrivial Burgers equation

$$(10.7) \qquad \partial_\tau q = (1 + \alpha\beta)\partial_{YY} q + (\beta - \alpha)\partial_Y(q^2)$$

for the evolution of the wave number q of the locally preferred planform.

Combining Theorem 3.4 with Theorem 10.2 gives the following result.

THEOREM 10.4. *For each fixed choice of integers $M \geq 1$, $0 < m < M$ and $n \geq M + 3$, and constants $C_0 > 0$ and $T_0 > 0$, there are constants $C_1 > 0$ and $\delta_1 > 0$ such that the following is true for each $\delta \in (0, \delta_1)$ and each solution q of the Burgers equation (10.7) for which*

$$\sup_{\tau\in[0,\tau_0]} \|q(\tau)\|_{H^n_{\text{ul}}} \leq C_0.$$

There exist higher-order approximations (q_h, r_h) with

$$\sup_{T\in[0,T_0]} \sup_{X\in\mathbb{R}} \left[\left|r_h(X,T) + \frac{1}{2}(q(X,T)^2 + \alpha\partial_X q(X,T))\right| + |q_h(X,T) - q(X,T)|\right]$$
$$\leq C_1\delta$$

and numbers $\varepsilon_2 > 0$ and $C_2 > 0$ such that, for each choice of $\varepsilon \in (0, \varepsilon_2)$, there is a solution $U = U(x, z, t)$ of the Taylor–Couette problem in PRII and a global phase $\phi_0(t)$ with $|\phi_0(t)| \leq C_2$ so that

$$\sup_{t\in[0,\tau_0/(\varepsilon\delta)^2]} \sup_{x\in[-1/(\varepsilon\delta)^m, 1/(\varepsilon\delta)^m]} \bigg| U(x - \phi_0(t), t) -$$
$$\varepsilon\left[1 + \delta^2 r_h(\varepsilon\delta(x - c_g t), \varepsilon^2\delta^2 t)\right] \times$$
$$\times \exp\left(i\ell_c x + i\omega_c + i\varepsilon^2\beta t + i\int_0^{\varepsilon\delta(x-c_g t)} \delta q_h(\delta Y, \varepsilon^2\delta^2 t)\, dY\right) U_1(\ell_c, z) - \text{c.c.}\bigg|$$
$$\leq C_1\varepsilon\delta^{1+M-m} + C_2\varepsilon^2.$$

Proposition 2.3 shows, at least on the level of approximation of solutions of the Taylor–Couette problem by the Burgers equation as explained by the theorems stated in this section, that the phases of waves with the same wave number are mixed universally in the Taylor–Couette problem in the case PRII near onset.

Bibliography

[1] I.S. Aranson and L. Kramer. The world of the complex Ginzburg–Landau equation. *Rev. Mod. Phys.* **74** (2002) 99–143.

[2] G. van Baalen. Phase turbulence in the complex Ginzburg–Landau equation via Kuramoto–Sivashinsky phase dynamics. *Comm. Math. Phys.* **247** (2004) 613–654.

[3] A.J. Bernoff. Slowly varying fully nonlinear wavetrains in the Ginzburg–Landau equation. *Physica D* **30** (1988) 363–381.

[4] V.N. Biktashev. Diffusion of autowaves: Evolution equation for slowly varying autowaves. *Physica D* **40** (1989) 83–90.

[5] J. Bricmont and A. Kupiainen. Renormalization group and the Ginzburg–Landau equation. *Comm. Math. Phys.* **150** (1992) 193–208.

[6] J. Bricmont, A. Kupiainen and G. Lin. Renormalization group and asymptotics of solutions of nonlinear parabolic equations. *Comm. Pure Appl. Math.* **47** (1994) 893–922.

[7] P. Chossat and G. Iooss. *The Couette–Taylor problem*. Springer, 1994.

[8] P. Collet and J.-P. Eckmann. The time dependent amplitude equation for the Swift–Hohenberg problem. *Comm. Math. Phys.* **132** (1990) 139–153.

[9] K.L. Cooke and S.M. Verduyn Lunel. Distributional and small solutions for linear time-dependent delay equations. *Diff. Int. Eqns.* **6** (1993) 1101–1117.

[10] M.C. Cross and P.C. Hohenberg. Pattern formation outside of equilibrium. *Rev. Mod. Phys.* **65** (1993) 851–1090.

[11] M.C. Cross and A.C. Newell. Convection patterns in large aspect ratio systems. *Physica D* **10** (1984) 299–328.

[12] O. Diekmann, S.A. van Gils, S.M. Verduyn Lunel and H.-O. Walther. *Delay equations: Functional, complex, and nonlinear analysis*. Springer, 1995.

[13] A. Doelman. Travelling waves in the complex Ginzburg-Landau equation. *J. Nonlinear Sci.* **3** (1993) 225–266.

[14] S.-I. Ei. The motion of weakly interacting pulses in reaction-diffusion systems. *J. Dynam. Diff. Eqns.* **14** (2002) 85–137.

[15] C. Elphick, E. Meron and E.A. Spiegel. Patterns of propagating pulses. *SIAM J. Appl. Math.* **50** (1990) 490–503.

[16] E.G. Eszter. *Evans function analysis of the stability of periodic travelling wave solutions of the FitzHugh–Nagumo system*. PhD thesis, University of Massachusetts, Amherst, 1999.

[17] A.A. Golovin, B.J. Matkowsky, A. Bayliss and A.A. Nepomnyashchy. Coupled KS-CGL and coupled Burgers-CGL equations for flames governed by a sequential reaction. *Physica D* **129** (1999) 253–298.

[18] G. Iooss, A. Mielke and Y. Demay. Theory of steady Ginzburg–Landau equation in hydrodynamic stability problems. *Europ. J. Mech. B/Fluids* **8** (1989)

229–268.

[19] A. van Harten. On the validity of Ginzburg–Landau's equation. *J. Nonlinear Sci.* **1** (1991) 397–422.

[20] A. van Harten. Modulated modulation equations. In: "Proceedings of the IUTAM/ISIMM Symposium on Structure and Dynamics of Nonlinear Waves in Fluids (Hannover, 1994)". World Science Publishing (1995) 117–130.

[21] S. Hastings. On the existence of homoclinic and periodic orbits for the FitzHugh–Nagumo equations. *Quart. J. Math. Oxford Ser.* **27** (1976) 123–134.

[22] D. Henry. *Geometric theory of semilinear parabolic equations.* Lecture Notes in Mathematics **840**. Springer, 1981.

[23] L.N. Howard and N. Kopell. Slowly varying waves and shock structures in reaction-diffusion equations. *Studies Appl. Math.* **56** (1976/77) 95–145.

[24] C.K.R.T. Jones. Stability of the travelling wave solution of the FitzHugh-Nagumo system. *Trans. Amer. Math. Soc.* **286** (1984) 431–469.

[25] T. Kapitula. Stability of weak shocks in λ-ω systems. *Indiana Univ. Math. J.* **40** (1991) 1193–1219.

[26] T. Kapitula. On the nonlinear stability of plane waves for the Ginzburg–Landau equation. *Comm. Pure Appl. Math.* **47** (1994) 831–841.

[27] T. Kapitula. Existence and stability of singular heteroclinic orbits for the Ginzburg–Landau equation. *Nonlinearity* **9** (1996) 669–685.

[28] Y. Katznelson. *An introduction to harmonic analysis.* Dover Publications, 1976.

[29] K. Kirchgässner. Wave-solutions of reversible systems and applications. *J. Diff. Eqns.* **45** (1982) 113–127.

[30] L. Kramer and W. Zimmermann. On the Eckhaus instability for spatially periodic patterns. *Physica D* **16** (1985) 221–252.

[31] Y. Kuramoto. *Chemical oscillations, waves, and turbulence.* Springer, 1984.

[32] J. Lega. Phase diffusion and weak turbulence. In: "Dynamics and Bifurcation of Patterns in Dissipative Systems" (Eds.: G. Dangelmayr and I. Oprea). World Scientific (2004) 183–157.

[33] A. Lunardi. *Analytic semigroups and optimal regularity in parabolic problems.* Birkhäuser, 1995.

[34] P. Manneville. *Dissipative structures and weak turbulence.* Academic Press, 1990.

[35] Y. Masutomi and N. Nozaki. Derivation of non-isotropic phase equations from a general reaction-diffusion equation. *Physica D* **151** (2001) 44–60.

[36] B.J. Matkowsky and V.A. Volpert. Stability of plane wave solutions of complex Ginzburg–Landau equations. *Quart. Appl. Math.* **51** (1993) 265–281.

[37] I. Melbourne. Derivation of the time-dependent Ginzburg–Landau equation on the line. *J. Nonlinear Sci.* **8** (1998) 1–15.

[38] I. Melbourne. Steady-state bifurcation with Euclidean symmetry. *Trans. Amer. Math. Soc.* **351** (1999) 1575–1603.

[39] I. Melbourne and G. Schneider. Phase dynamics in the real Ginzburg–Landau equation. *Math. Nachr.* **263/264** (2004) 171–180.

[40] I. Melbourne and G. Schneider. Phase dynamics in the complex Ginzburg–Landau equation. *J. Diff. Eqns.* **199** (2004) 22–46.

[41] A. Mielke. A spatial center manifold approach to steady bifurcations from spatially periodic patterns. In: "Dynamics in Dissipative Systems: Reductions,

Bifurcations and Stability" (Eds.: G. Dangelmayr, B. Fiedler, K. Kirchgässner and A. Mielke). Pitman Research Notes **352** (1996) 209–262.

[42] A. Mielke. The Ginzburg–Landau equation in its role as a modulation equation. In: "Handbook of Dynamical Systems II" (Ed.: B. Fiedler). North-Holland (2002) 759–834.

[43] H. Mori and Y. Kuramoto. *Dissipative structures and chaos*. Springer, 1998.

[44] A. Newell, T. Passot and J. Lega. Order parameter equations for patterns. *Annu. Rev. Fluid Mech.* **25** (1993) 399–453.

[45] T. Passot and A.C. Newell. Towards a universal theory for natural patterns. *Physica D* **74** (1994) 301–352.

[46] M. Reed and B. Simon. *Methods of modern mathematical physics 1–4*. Academic Press, 1972.

[47] B. Sandstede. Weak interaction of pulses. In preparation.

[48] B. Sandstede and A. Scheel. On the stability of periodic travelling waves with large spatial period. *J. Diff. Eqns.* **172** (2001) 134–188.

[49] B. Sandstede and A. Scheel. On the structure of spectra of modulated travelling waves. *Math. Nachr.* **232** (2001) 39–93.

[50] B. Sandstede and A. Scheel. Defects in oscillatory media: toward a classification. *SIAM J. Appl. Dynam. Syst.* **3** (2004) 1–68.

[51] B. Scarpellini. *Stability, instability, and direct integrals*. Pitman Research Notes **402**. Chapman & Hall, 1999.

[52] G. Schneider. Error estimates for the Ginzburg–Landau approximation. *Z. Angew. Math. Physik* **45** (1994) 433–457.

[53] G. Schneider. Validity and limitation of the Newell–Whitehead equation. *Math. Nachr.* **176** (1995) 249–263.

[54] G. Schneider. Nonlinear stability of Taylor vortices in infinite cylinders. *Arch. Ration. Mech. Anal.* **144** (1998) 121–200.

[55] G. Schneider. Global existence results for pattern forming processes in infinite cylindrical domains—Applications to 3D Navier–Stokes problems. *J. Math. Pures Appl.* **78** (1999) 265–312.

[56] G.B. Whitham. *Linear and nonlinear waves*. Wiley, 1974.

[57] E. Yanagida. Stability of fast travelling pulse solutions of the FitzHugh–Nagumo equations. *J. Math. Biology* **22** (1985) 81–104.

[58] S. Zelik and A. Mielke. Multi-pulse evolution and space-time chaos in dissipative systems. *Memoirs Amer. Math. Soc.* (to appear).

Editorial Information

To be published in the *Memoirs*, a paper must be correct, new, nontrivial, and significant. Further, it must be well written and of interest to a substantial number of mathematicians. Piecemeal results, such as an inconclusive step toward an unproved major theorem or a minor variation on a known result, are in general not acceptable for publication.

Papers appearing in *Memoirs* are generally at least 80 and not more than 200 published pages in length. Papers less than 80 or more than 200 published pages require the approval of the Managing Editor of the Transactions/Memoirs Editorial Board.

As of January 31, 2009, the backlog for this journal was approximately 11 volumes. This estimate is the result of dividing the number of manuscripts for this journal in the Providence office that have not yet gone to the printer on the above date by the average number of monographs per volume over the previous twelve months, reduced by the number of volumes published in four months (the time necessary for preparing a volume for the printer). (There are 6 volumes per year, each usually containing at least 4 numbers.)

A Consent to Publish and Copyright Agreement is required before a paper will be published in the *Memoirs*. After a paper is accepted for publication, the Providence office will send a Consent to Publish and Copyright Agreement to all authors of the paper. By submitting a paper to the *Memoirs*, authors certify that the results have not been submitted to nor are they under consideration for publication by another journal, conference proceedings, or similar publication.

Information for Authors

Memoirs are printed from camera copy fully prepared by the author. This means that the finished book will look exactly like the copy submitted.

Initial submission. The AMS uses Centralized Manuscript Processing for initial submissions. Authors should submit a PDF file using the Initial Manuscript Submission form found at www.ams.org/peer-review-submission, or send one copy of the manuscript to the following address: Centralized Manuscript Processing, MEMOIRS OF THE AMS, 201 Charles Street, Providence, RI 02904-2294 USA. If a paper copy is being forwarded to the AMS, indicate that it is for it Memoirs and include the name of the corresponding author, contact information such as email address or mailing address, and the name of an appropriate Editor to review the paper (see the list of Editors below).

The paper must contain a *descriptive title* and an *abstract* that summarizes the article in language suitable for workers in the general field (algebra, analysis, etc.). The *descriptive title* should be short, but informative; useless or vague phrases such as "some remarks about" or "concerning" should be avoided. The *abstract* should be at least one complete sentence, and at most 300 words. Included with the footnotes to the paper should be the 2000 *Mathematics Subject Classification* representing the primary and secondary subjects of the article. The classifications are accessible from www.ams.org/msc/. The list of classifications is also available in print starting with the 1999 annual index of *Mathematical Reviews*. The Mathematics Subject Classification footnote may be followed by a list of *key words and phrases* describing the subject matter of the article and taken from it. Journal abbreviations used in bibliographies are listed in the latest *Mathematical Reviews* annual index. The series abbreviations are also accessible from www.ams.org/msnhtml/serials.pdf. To help in preparing and verifying references, the AMS offers MR Lookup, a Reference Tool for Linking, at www.ams.org/mrlookup/.

Electronically prepared manuscripts. The AMS encourages electronically prepared manuscripts, with a strong preference for $\mathcal{A}_{\mathcal{M}}\mathcal{S}$-LaTeX. To this end, the Society has prepared $\mathcal{A}_{\mathcal{M}}\mathcal{S}$-LaTeX author packages for each AMS publication. Author packages include instructions for preparing electronic manuscripts, samples, and a style file that generates the particular design specifications of that publication series. Though $\mathcal{AMS}$-LaTeX is the highly preferred format of TeX, author packages are also available in $\mathcal{AMS}$-TeX.

Authors may retrieve an author package for *Memoirs of the AMS* from www.ams.org/journals/memo/memoauthorpac.html or via FTP to ftp.ams.org (login as anonymous, enter username as password, and type cd pub/author-info). The *AMS Author Handbook* and the *Instruction Manual* are available in PDF format from the author package link. The author package can also be obtained free of charge by sending email to tech-support@ams.org (Internet) or from the Publication Division, American Mathematical Society, 201 Charles St., Providence, RI 02904-2294, USA. When requesting an author package, please specify $\mathcal{AMS}$-LaTeX or $\mathcal{AMS}$-TeX and the publication in which your paper will appear. Please be sure to include your complete mailing address.

After acceptance. The final version of the electronic file should be sent to the Providence office (this includes any TeX source file, any graphics files, and the DVI or PostScript file) immediately after the paper has been accepted for publication.

Before sending the source file, be sure you have proofread your paper carefully. The files you send must be the EXACT files used to generate the proof copy that was accepted for publication. For all publications, authors are required to send a printed copy of their paper, which exactly matches the copy approved for publication, along with any graphics that will appear in the paper.

Accepted electronically prepared files can be submitted via the web at www.ams.org/submit-book-journal/, sent via FTP, or sent on CD-Rom or diskette to the Electronic Prepress Department, American Mathematical Society, 201 Charles Street, Providence, RI 02904-2294 USA. TeX source files, DVI files, and PostScript files can be transferred over the Internet by FTP to the Internet node ftp.ams.org (130.44.1.100). When sending a manuscript electronically via CD-Rom or diskette, please be sure to include a message identifying the paper as a Memoir.

Electronically prepared manuscripts can also be sent via email to pub-submit@ams.org (Internet). In order to send files via email, they must be encoded properly. (DVI files are binary and PostScript files tend to be very large.)

Electronic graphics. Comprehensive instructions on preparing graphics are available at www.ams.org/authors/journals.html. A few of the major requirements are given here.

Submit files for graphics as EPS (Encapsulated PostScript) files. This includes graphics originated via a graphics application as well as scanned photographs or other computer-generated images. If this is not possible, TIFF files are acceptable as long as they can be opened in Adobe Photoshop or Illustrator. No matter what method was used to produce the graphic, it is necessary to provide a paper copy to the AMS.

Authors using graphics packages for the creation of electronic art should also avoid the use of any lines thinner than 0.5 points in width. Many graphics packages allow the user to specify a "hairline" for a very thin line. Hairlines often look acceptable when proofed on a typical laser printer. However, when produced on a high-resolution laser imagesetter, hairlines become nearly invisible and will be lost entirely in the final printing process.

Screens should be set to values between 15% and 85%. Screens which fall outside of this range are too light or too dark to print correctly. Variations of screens within a graphic should be no less than 10%.

Inquiries. Any inquiries concerning a paper that has been accepted for publication should be sent to memo-query@ams.org or directly to the Electronic Prepress Department, American Mathematical Society, 201 Charles St., Providence, RI 02904-2294 USA.

Editors

This journal is designed particularly for long research papers, normally at least 80 pages in length, and groups of cognate papers in pure and applied mathematics. Papers intended for publication in the *Memoirs* should be addressed to one of the following editors. The AMS uses Centralized Manuscript Processing for initial submissions to AMS journals. Authors should follow instructions listed on the Initial Submission page found at www.ams.org/memo/memosubmit.html.

Algebra to ALEXANDER KLESHCHEV, Department of Mathematics, University of Oregon, Eugene, OR 97403-1222; email: ams@noether.uoregon.edu

Algebraic geometry to DAN ABRAMOVICH, Department of Mathematics, Brown University, Box 1917, Providence, RI 02912; email: amsedit@math.brown.edu

Algebraic geometry and its applications to MINA TEICHER, Emmy Noether Research Institute for Mathematics, Bar-Ilan University, Ramat-Gan 52900, Israel; email: teicher@macs.biu.ac.il

Algebraic topology to ALEJANDRO ADEM, Department of Mathematics, University of British Columbia, Room 121, 1984 Mathematics Road, Vancouver, British Columbia, Canada V6T 1Z2; email: adem@math.ubc.ca

Combinatorics to JOHN R. STEMBRIDGE, Department of Mathematics, University of Michigan, Ann Arbor, Michigan 48109-1109; email: JRS@umich.edu

Commutative and homological algebra to LUCHEZAR L. AVRAMOV, Department of Mathematics, University of Nebraska, Lincoln, NE 68588-0130; email: avramov@math.unl.edu

Complex analysis and harmonic analysis to ALEXANDER NAGEL, Department of Mathematics, University of Wisconsin, 480 Lincoln Drive, Madison, WI 53706-1313; email: nagel@math.wisc.edu

Differential geometry and global analysis to CHRIS WOODWARD, Department of Mathematics, Rutgers University, 110 Frelinghuysen Road, Piscataway, NJ 08854; email: ctw@math.rutgers.edu

Dynamical systems and ergodic theory and complex analysis to YUNPING JIANG, Department of Mathematics, CUNY Queens College and Graduate Center, 65-30 Kissena Blvd., Flushing, NY 11367; email:CcCC Yunping.Jiang@qc.cuny.edu

Functional analysis and operator algebras to DIMITRI SHLYAKHTENKO, Department of Mathematics, University of California, Los Angeles, CA 90095; email: shlyakht@math.ucla.edu

Geometric analysis to WILLIAM P. MINICOZZI II, Department of Mathematics, Johns Hopkins University, 3400 N. Charles St., Baltimore, MD 21218; email: trans@math.jhu.edu

Geometric topology to MARK FEIGHN, Math Department, Rutgers University, Newark, NJ 07102; email: feighn@andromeda.rutgers.edu

Harmonic analysis, representation theory, and Lie theory to ROBERT J. STANTON, Department of Mathematics, The Ohio State University, 231 West 18th Avenue, Columbus, OH 43210-1174; email: stanton@math.ohio-state.edu

Logic to STEFFEN LEMPP, Department of Mathematics, University of Wisconsin, 480 Lincoln Drive, Madison, Wisconsin 53706-1388; email: lempp@math.wisc.edu

Number theory to JONATHAN ROGAWSKI, Department of Mathematics, University of California, Los Angeles, CA 90095; email: jonr@math.ucla.edu

Number theory to SHANKAR SEN, Department of Mathematics, 505 Malott Hall, Cornell University, Ithaca, NY 14853; email: ss70@cornell.edu

Partial differential equations to GUSTAVO PONCE, Department of Mathematics, South Hall, Room 6607, University of California, Santa Barbara, CA 93106; email: ponce@math.ucsb.edu

Partial differential equations and dynamical systems to PETER POLACIK, School of Mathematics, University of Minnesota, Minneapolis, MN 55455; email: polacik@math.umn.edu

Probability and statistics to RICHARD BASS, Department of Mathematics, University of Connecticut, Storrs, CT 06269-3009; email: bass@math.uconn.edu

Real analysis and partial differential equations to DANIEL TATARU, Department of Mathematics, University of California, Berkeley, Berkeley, CA 94720; email: tataru@math.berkeley.edu

All other communications to the editors should be addressed to the Managing Editor, ROBERT GURALNICK, Department of Mathematics, University of Southern California, Los Angeles, CA 90089-1113; email: guralnic@math.usc.edu.

Titles in This Series

935 **Mihai Ciucu,** The scaling limit of the correlation of holes on the triangular lattice with periodic boundary conditions, 2009

934 **Arjen Doelman, Björn Sandstede, Arnd Scheel, and Guido Schneider,** The dynamics of modulated wave trains, 2009

933 **Luchezar Stoyanov,** Scattering resonances for several small convex bodies and the Lax-Phillips conjecture, 2009

932 **Jun Kigami,** Volume doubling measures and heat kernel estimates on self-similar sets, 2009

931 **Robert C. Dalang and Marta Sanz-Solé,** Hölder-Sobolev regularity of the solution to the stochastic wave equation in dimension three, 2009

930 **Volkmar Liebscher,** Random sets and invariants for (type II) continuous tensor product systems of Hilbert spaces, 2009

929 **Richard F. Bass, Xia Chen, and Jay Rosen,** Moderate deviations for the range of planar random walks, 2009

928 **Ulrich Bunke,** Index theory, eta forms, and Deligne cohomology, 2009

927 **N. Chernov and D. Dolgopyat,** Brownian Brownian motion-I, 2009

926 **Riccardo Benedetti and Francesco Bonsante,** Canonical wick rotations in 3-dimensional gravity, 2009

925 **Sergey Zelik and Alexander Mielke,** Multi-pulse evolution and space-time chaos in dissipative systems, 2009

924 **Pierre-Emmanuel Caprace,** "Abstract" homomorphisms of split Kac-Moody groups, 2009

923 **Michael Jöllenbeck and Volkmar Welker,** Minimal resolutions via algebraic discrete Morse theory, 2009

922 **Ph. Barbe and W. P. McCormick,** Asymptotic expansions for infinite weighted convolutions of heavy tail distributions and applications, 2009

921 **Thomas Lehmkuhl,** Compactification of the Drinfeld modular surfaces, 2009

920 **Georgia Benkart, Thomas Gregory, and Alexander Premet,** The recognition theorem for graded Lie algebras in prime characteristic, 2009

919 **Roelof W. Bruggeman and Roberto J. Miatello,** Sum formula for SL_2 over a totally real number field, 2009

918 **Jonathan Brundan and Alexander Kleshchev,** Representations of shifted Yangians and finite W-algebras, 2008

917 **Salah-Eldin A. Mohammed, Tusheng Zhang, and Huaizhong Zhao,** The stable manifold theorem for semilinear stochastic evolution equations and stochastic partial differential equations, 2008

916 **Yoshikata Kida,** The mapping class group from the viewpoint of measure equivalence theory, 2008

915 **Sergiu Aizicovici, Nikolaos S. Papageorgiou, and Vasile Staicu,** Degree theory for operators of monotone type and nonlinear elliptic equations with inequality constraints, 2008

914 **E. Shargorodsky and J. F. Toland,** Bernoulli free-boundary problems, 2008

913 **Ethan Akin, Joseph Auslander, and Eli Glasner,** The topological dynamics of Ellis actions, 2008

912 **Igor Chueshov and Irena Lasiecka,** Long-time behavior of second order evolution equations with nonlinear damping, 2008

911 **John Locker,** Eigenvalues and completeness for regular and simply irregular two-point differential operators, 2008

910 **Joel Friedman,** A proof of Alon's second eigenvalue conjecture and related problems, 2008

TITLES IN THIS SERIES

909 **Cameron McA. Gordon and Ying-Qing Wu,** Toroidal Dehn fillings on hyperbolic 3-manifolds, 2008

908 **J.-L. Waldspurger,** L'endoscopie tordue n'est pas si tordue, 2008

907 **Yuanhua Wang and Fei Xu,** Spinor genera in characteristic 2, 2008

906 **Raphaël S. Ponge,** Heisenberg calculus and spectral theory of hypoelliptic operators on Heisenberg manifolds, 2008

905 **Dominic Verity,** Complicial sets characterising the simplicial nerves of strict ω-categories, 2008

904 **William M. Goldman and Eugene Z. Xia,** Rank one Higgs bundles and representations of fundamental groups of Riemann surfaces, 2008

903 **Gail Letzter,** Invariant differential operators for quantum symmetric spaces, 2008

902 **Bertrand Toën and Gabriele Vezzosi,** Homotopical algebraic geometry II: Geometric stacks and applications, 2008

901 **Ron Donagi and Tony Pantev (with an appendix by Dmitry Arinkin),** Torus fibrations, gerbes, and duality, 2008

900 **Wolfgang Bertram,** Differential geometry, Lie groups and symmetric spaces over general base fields and rings, 2008

899 **Piotr Hajłasz, Tadeusz Iwaniec, Jan Malý, and Jani Onninen,** Weakly differentiable mappings between manifolds, 2008

898 **John Rognes,** Galois extensions of structured ring spectra/Stably dualizable groups, 2008

897 **Michael I. Ganzburg,** Limit theorems of polynomial approximation with exponential weights, 2008

896 **Michael Kapovich, Bernhard Leeb, and John J. Millson,** The generalized triangle inequalities in symmetric spaces and buildings with applications to algebra, 2008

895 **Steffen Roch,** Finite sections of band-dominated operators, 2008

894 **Martin Dindoš,** Hardy spaces and potential theory on C^1 domains in Riemannian manifolds, 2008

893 **Tadeusz Iwaniec and Gaven Martin,** The Beltrami Equation, 2008

892 **Jim Agler, John Harland, and Benjamin J. Raphael,** Classical function theory, operator dilation theory, and machine computation on multiply-connected domains, 2008

891 **John H. Hubbard and Peter Papadopol,** Newton's method applied to two quadratic equations in $\mathbb{C}^2$ viewed as a global dynamical system, 2008

890 **Steven Dale Cutkosky,** Toroidalization of dominant morphisms of 3-folds, 2007

889 **Michael Sever,** Distribution solutions of nonlinear systems of conservation laws, 2007

888 **Roger Chalkley,** Basic global relative invariants for nonlinear differential equations, 2007

887 **Charlotte Wahl,** Noncommutative Maslov index and eta-forms, 2007

886 **Robert M. Guralnick and John Shareshian,** Symmetric and alternating groups as monodromy groups of Riemann surfaces I: Generic covers and covers with many branch points, 2007

885 **Jae Choon Cha,** The structure of the rational concordance group of knots, 2007

884 **Dan Haran, Moshe Jarden, and Florian Pop,** Projective group structures as absolute Galois structures with block approximation, 2007

883 **Apostolos Beligiannis and Idun Reiten,** Homological and homotopical aspects of torsion theories, 2007

For a complete list of titles in this series, visit the
AMS Bookstore at **www.ams.org/bookstore/**.